REAL WEST BALTIMORE STORY
— — STORY —

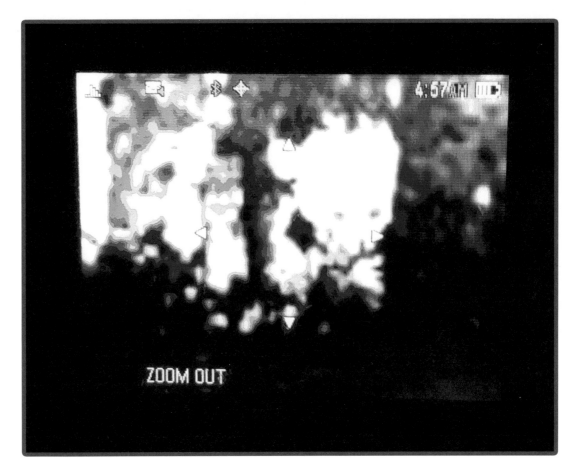

Get the information that the government
doesn't want you to know.

REAL UFOS AND ALIEN BEINGS
IN THE HOOD

T R U T H

AuthorHouse™
1663 Liberty Drive
Bloomington, IN 47403
www.authorhouse.com
Phone: 1 (800) 839-8640

Published by AuthorHouse 12/13/2018

ISBN: 978-1-5462-7125-3 (sc)
ISBN: 978-1-5462-7126-0 (e)

Print information available on the last page.

This book is printed on acid-free paper.

authorHOUSE®

REAL WEST

BALTIMORE

—STORY—

This book is dedicated to Mr.&Mrs Emma and George B., Mr.&Mrs. George and Olive W., Mr.&Mrs. Bruce and Liz C., Mr.&Mrs. George and Ruby B., Albert R., Pauline T., and to my parents Mr.&Mrs. A and G T. Also I wanted to send out love and blessing to my comrades Coakley, Michel Lane, Lil Marvin, Fat Larry, Todd, Troy, Flip, Allan Brown, OOG, Lil Magic, Lil H, Adof Hitler, Jrock, C & B, Leroy H, Kev T, Bobboy TJ, TJT, Lee Bo, DP, OG, Lil Legs, OOG Kirk, Eric Trusedale, Fud, Pud, BC.

CONTENTS

INTRODUCTION

This book depicts UFO'S in the hood. Maryland has been colonized by alien beings, period.

CHAPTER 1

95% of this story is true, beginning with me at the age of about 4 years of age. Living in our West Eglestone home, I can remember a lady dressed in all white visiting my bedroom on a regular, at night. I'm not sure if she was an alien, but I really do believe I know they watched over of me now and have been for quite sometime.

The lady in white never tried to harm me, but I felt she was there to protect me from something at the time. I wasn't sure at the time from what. We stayed at our home until around 1976. My guardian angel would visit me often. I was afraid and felt safe at the same time. Toward the end of 1976, our home vandalized and our beautiful german shepherd, Peppi, was poisoned and died. We moved to an avenue in the heart of the hood. A year later, we moved a little furthur uptown to a better neighborhood and a larger home. About six months into moving into our new home, all sorts of things started to happen, including my very first UFO sighting. I learned a lot at a very early age. Both parents worked hard and retired from respectable companies. They weren't the kind of parents to run the streets and leave us unattended.

My older brother and I had very different ideas about our fate growing up in the hood. In our early days, we played scully king of the hills, hot butter bean, baseball, football, and boxing.

I joined the Karate Crew who's office somewhere in park nearby would come into play to help me survive even until this day. I was one of the best wrestlers, boxers, and ball players around for my size. I attended Vine Elementary School for kindergarten, and Catholic School for the first through the fifth grade. The grades I received were good, nuns ain't play and those schools got more out of me than I ever knew I had to give. O.K. By now its 1980 and my aunt has relocated from a village to another place. Now, keep in mind, the very weekend she moves in, she throws a moving cookout. The entire family comes out to check out her new crib. Now, while there, my little cousin, and my longtime homegirl, were out in the backyard, the adults still inside, the three of us decided to go outside the boundaries of the yard, and out into the alley behind the house. We weren't out there a good ten minutes, before me and Tee, she was 6 years old and I was 8 and E was only 2, we all saw a bright glowing object in the sky in broad daylight. I knew being the oldest of the three, knew this was not a normal object. We saw it, me and her, looked at each other, and asked, "did you see that?" We both said, "yea" and continued to stare as the object hovered. The next thing I knew, the three of us were walking back up to the house, even though I knew we were just in the alley! At this point, the sky looked different. It was as if three hours had passed in a matter of minutes. Remember, at the time of this occurence, I was eight years old, so I didn't put it all together, missing time, that is. This actually happened, and this would be the first of many sightings.

Now, while attending Catholic school, I would walk to and from school. My home was only about eight blocks from the school. I can remember almost getting hit by a car, right in front of my home. I fell out of the back seat of a moving car and didn't get a scratch! I came very close to being abducted somewhere in the city fair by an ice ream man of all people! He tried to lure me in the back of his truck in order to get some free ice cream. I didn't fall for his game, and went in the opposite direction. My guardian angel sitting on my shoulder! !

The neighborhood bad boys got into the habit of cutting across our yard in order to get into the next block. This habit didn't go well with me and my crew. We had fights every morning before heading off to school. I decided enough was enough, so I obtained a 22 revolver and made the mistake of taking it to school. The gun was reported and I was suspended. Needless to say, my parents was none to happy about the matter! The Edgewood boys heard about the gun and never again crossed our yard in a disrespectful manner!

By the time I reached third grade, I was reading on a six grade level to 8th grade level. The Nuns would come by our home after school just to hear me read. My mom always read to me a lot. Now my math, not so good at least then. The fifth grade was my last grade at the Catholic School. I graduated from a Catholic school and then attended Knight Elementary School. I was placed in Ms. Stokes class, second smartest there. Five years later, we moved to Ville. I met a slew of people who would influence my life forever. I always associated myself with older dudes.

The guys I met always thought of me as family. At the tender age of 10, I started stealing, fighting, and rebelling against my mom and dads house rules. The dudes I was hanging out with were straight up gangters. I had homies who would show up at schools on any given day and terrorize everyone. Even I was terrorized for a short time as well. If you were new to the hood, they didn't except you at first, and sometimes at all, meaning death by association. So, it was rough, but little did I know I would have to be perfect to survive the streets, and almost was. Once on Ville, I started to grow up and see what was within my reach. I passed the 6th grade and being fresh, and having a girl, was important to my image, so I found ways to keep fresh and kept a girl. As my fighting improved, my image grew. I fought a lot, like everyday dudes bigger, stronger, and most of the time older. No one would boaster about locking ass with me. My older brother was known for fighting too.

As time passed, me and my crew got closer. We did everything together, we were never separated. The older hustlers loved our unity. We all respected each other for what we did in the hood. Now our block was the truth and had been for awhile, meaning, money, woman and respect, that's it! We started getting money at age 13, me personally, I was 12. My original partner in crime, Rahoo, got connected early, and even being as young as we were, we kept that money straight. We got a good rep for being young bloods or young hoodlums who knew what they were doing in the eyes of the old heads. Since we had potential and family for getting money the legal way, we already had the DNA to make it big on the streets. Our parents believe in making an honest living way before we were even born.

Now, after seeing the UFO's sightings, I believed doing right meant less to me for some reason. Maybe they implanted me at the age of eight to cause me a life of excitement and fulfillment. Me being so into the subject I believe is why they have chosen me. I mean I've seen UFO's aliens, ghosts, who even appeared to me in the brood daylight. Yeah!

Right in the hood. Now we started selling and hustling on Saverly and Middletown. We started out selling raw coke for an old head gangster by the name of K.P. He pulled me and Rahod in to hold while he goes on the block to send the sales our way. We had to hold the gun, which I offered to do at the time knowing I was a juvenile, and if I got caught, maybe probation, but no jail time for a first offense, juvenile or adult in most cases as long as you were represented by an attorney. I learned at an early age, if you are hustling you need a lawyer on standby, and surprising enough we had two. So anyway, we're around the apts. everyday. We would sell out everyday and we took in about $1000.00 nicks a day at first.We just grabed them out of plastic bags and distributed them. Viles were already packed with cut cocaine. There weren't many cops around at this time, so we thrived. We would make about $50.00 a week each at first. Hey! Were you making that kind of money at 13 years of age? $50.00 a week was great! Throughout this time, we were gaining respect all around the table because they now knew we weren't afraid to hold heat or go to jail. Trust this, we're in and have been ever since,

We are both in our fourties now, and we're still alive, go figure! So now we are both attending High School for the 7 through 9th grades. My class of 903 was the last class to graduate from that school. My friends were not into boxing for cash yet, but all they did then was knock niggers dicks strings out that got out of line. That's right the man, the undisputed heavy weight champ went to junior high school with me and all they all know Lol u-u from Saverly and Middletown.

You see, even before I started getting small money, I wanted for nothing. My older brother Dirt, kept me fresh. Dirt worked for an athletic clothing and footwear store. He could get any tennis shoe, sweat suit, hat, and great gear, if they sold them, he had access to them. Dirt made thousands of dollars just selling the tennis shoes at half price. Yeah, niggers knew we got down like that! Our entire hood did not start to get kilo money til we turned to ready rock. Oh my God, now the whole time we getting money, we are still seeing UFO's at different times as well. So now, it's Mugs, Iopp, Rahoo and me. Sparky and Von came up later for real. But, anyway, we pleaded nothing would come between us as we embarked on this journey to get money, and most of us still feel that way even unto this day. The old head was selling dope in the Ville Court, while we handled the ready and raw.

Spotting UFO's at night was easy because we shot Lisnt's out or they would already off because the city didn't replace the burned out lights on the regular. We would see starlight objects moving as if they were on a cosmic highway. One night, we had about thirty of them over head moving slowly, and we all saw them. So, yes, UFO's were in our hood as I became a man on the streets of Eglestone. My home did not believe in UFO's ass I do, I don't think. We have never discussed them other then we knew they existed.

As we got our clientele in order while skipping school, we also had to deal with fights, shoot-outs and home drama. We kept force while we had 24-7 shop and was bringing in hundreds of thousands a year, probably in total profits just off our block. You see, we were the first niggers to sell ready rock in Eglestone period. That's the truth, so yeah, my G's that's our title. I don't believe anyone in the city, even to this day, made that type of money on a 15 year run. Going to jail a few times along the way, really was part of the plan. I would return home, start the process all over again, never skipping a beat, and we wanted for nothing, the city was ours!

CHAPTER 2

Still until this day we have strength in numbers! We all were good at what we did, all soldiers on the front line, but none like little Du Du, okay, and I took my street mentality to another level once I got serious about cash, we all did! We had old heads behind us, young guys, guys our age, and ones that could not wait to get paid. We were called the bottom boys, readdy boys, Saverly and Middletown crew, but the bottom line was we were all street level dealers who made money daily. Fuck the law, stick up boys, spending time with women, all that. Just cash and fast! I stayed on the block daily, for hours at a time, only leaving if I sold out, or got locked-up. Trust me on most days, I sold out at will, on the bundles of ready.

To give ya'll some type of perspective of the cash, by 1990 1 was living in More County. Now, back then I was paying about $900.00 a month for rent, gas & electric, cable, plus I had a $90.00 habit daily of buying herb for myself and for my girl. My habit alone, and I can't even tell you how much that cost me in a year. I started smoking weed at the age of 14 and Roxbury were my cigarette of choice. So, lets say $90.00 a day for weed is $630.00 a week, so you can imagine I made well over a mil, so yeah, we were getting it for sure! I;m sure I spent at least $30,000 or more a year on urb alone, I'm not sure you think maybe it was the weed messing with my memory? So, I would say on any given day, once we all established clout on the block you could easily have 15 different color vals at $10,000 a piece and everybody sold out at our peak, the math speaks for itself, $10,000 times 15=$150.00 on our block at one time!

Real talk people, so you can see a million was made on our strip many times over.

Even before we started getting money, G's were out there before we were Hustling was a way of life, but so was violence and chaos. Now on our come up we beef with several other notorious hoods, our West side was the most deadly around. Gun play, yes, sometimes, but back then, you can beat a nigger up and not worry sometimes. When gun play was involved, well, that was an entirely different scenario! Young and dumb, and of course, you know the rest. But, we had respect, cars, women, clothes, Tims, and jewelry, I had my own crib as well.

Now, we were close, very close, so we knew about each others goods, bads, everything at one time. There were several OG's whom my mob looked up to in the early days. These niggers seemed to have it all, and our backs as well. We were well connected and still are, but for the record, I'm retired, so I'm gonna keep it real! Bob, Allen, A.D., Lil Kirk, Cabo,Lil Remy, Lil Leg, Joey Brinkley, Mike Brinkley, Lil Jeff, Lil Dana, Tank, Pookie

Shaw, Lil Jimmy, Kevin P., Fat Larry, Dante, K.J., Adolf, AKA, Maddog, Hitler, Nut, Michael L, Lil Toby, Ham, Pokey, Nut, David H., Coakley, Lil Rodney, Troy, Flip, Pie, Tiny, Yvette, Tay, Mugs, loop, Fat Derrick, Lil Ricky, Sammy, Duane, Eurkie, Lon, Black, Stenton, Cassandra, Donna, Tisha, Kee Kee, Michelle, Joy, Lil Joe, Lil Joe, Lil Mal, Gregg, J. Hines, Lil Nanny, Fat Paul, Cornelias, Harvey, Raon, Razor Rod, Stan Dog, Nick, Nat, Ms. Bawanda, Ms.Nicey, Aunt Teresa, Samy, Keith H, Dante, Black Eric, Fred, Moe, Poopie, Weenie, Von, Pig, Butchie, Bandit, Malik, J.T, Pop, Dirt, Rue, Uncle B, Uncle B, Uncle Punnell, Uncle David, okay, that's enough. All these peeps touched my life at one time or another and a lot of them are dead and gone, but I say may God bless them all!I love you all and this experience and if only I knew then, what I know now, maybe some of them would have survived. You know the streets and believe me, UFO's were seen during trying times in the hood, at least in hood. I was not the only one whom had seen them!

So, with the money came bullshit, jail, bail, lawyers, probation, parole, judges, courts, and of course money. We knew the rules before we joined the business. Staying true was the good, but was hard to do for everyone, but somehow, I did it. Yes, I made mistakes, but you never disrespected a OG who came before you. You don't disrepect them because you don't know what they went through before your time. Right there, on the same block where you reside, you see money! Stay true to your crew, no matter what! Follow the rules and stick to them. Now UFO's seemed to show up in my life during difficult times. While I hustled to pay rent, and stay fresh, and hip and cool all round the table. It was important in my hood to represent well. I was lucky I had all that strength and honesty as well.

For instance, I've locked ass with a lot of dudes coming up mostly fighting, but a test of strength and I've lost in a real situation only once. So, I would say my lock ass record on this planet would be around 60-1. Now, real fighting I'm 150 pds. Know on the streets of West B-More. That's real life and destruction too! The lock ass lost, no punches were ever thrown and the big ass nigger went to the hospital and got 6 stitches, I in return, was out about 6 seconds as well. We were working hard and got into a verbal altercation that turned violent, quick. After I came to, I hugged him and he knew he had won! He now had to respect my G period and that's how I've lived my life. Respect first, and everything else falls in place.

When you haved it, and I earned it the hard way, and I loved every minute of it as well.

Some won't believe this story after reading it, but believe me its 95.5% true. If your wondering what happened in the lock ass contest, we were both trippin. I got tired of talking and walked down on him fast in about 3.9 astro seconds. I grabbed him, plus, he was about 5'8", 225 pds., I stood about 5'10, 153pds. I wanted to pick him up and slam him, but I couldn't.

His sweaty ass standing chest to chest, he says to me," Come on shorty, you can't get me like that," so I grabbed a little tighter, slid down his legs, and in the process when I scooped him up in the air and slammed him down, I was on top of him, but my head went up under his shirt, and when my bare face brushed up against his dirty sweaty stinky chest, I almost passed out! He grabbed his shirt, with my head still inside, and held on for dear life. He stunk so bad, I passed out in 2 milli seconds. Thank God he didn't know I was out and couldn't take advantage of the situation. In the seconds I was out, he got away from me. When I looked up, my knife was in the dirt. I usually kept two knives on me after my conviction in 91. This incident took place in 07. When I slammed him to the ground on a second time, a piece of metal caught him in his back, and thats how he came to need six stiches.

In 1999, facing three felonies at the same time, for the second time in my life. I went to jail. After my three year stay, I came home and never sold cocaine as of yet. Yeah I saw UFO's before and after that incident! I have never really lost a contact situation other to my right hand man Sam Cook. He was like a brother. Sam Cook was better than me in a number of ways. Now, please don't think that I'm going in a lot of directions, but I promise, its all going to come together, ok. UFO's may control a lot of human beings who have been implanted from a distance, and I think they watched us on the regular. Just as we watch reality shows, only for them its us! This is what I believe, the technology, I know we don't have yet on their level. If they are working with the governments small versions. I believe just what they are capable of. They must need us, they are into torturing poor human beings before killing us, putting some people through horrific problems making worse case scenarios of thought come true. At least, that's what happened to me.

Several times in my life, even though I knew it was going to happen, I could not stop the events, real shit too! The death of my father, Gary, and my grandfather, Albert, my Aunt Alease and lots of uncle's, died from everything from murder to pancreatic cancer. My best friend was my Uncle Kevin, he was the closest uncle to me. When he died of pancreatic cancer, no amount of smoke or drink could ease the pain of his loss. Kevin is the reason I'm writing this book. Shits deep y'all, be aware that there are beings out here that can touch our lives, and as a matter of fact, I believe the came from very far away to do so. I've been investigating UFO's for over 34 yrs. Real talk, ever since my first sighting in the summer of 1980.

I'm so glad that Third Phase of the moon took the time to make a very big different in the world of UFO's and alien contacts. The E.T.'s seem to like many of the same things that we humans enjoy. Things like sex and violence, what a coincidence, huh? Well anyway, I'm going to jump back and forth and hopefully, it will be published that way. This will be my first book of many to come if people support the truth I don't believe in coincidences in life, everything happens for a reason, big, small, whatever, and it all comes together in the right situation. Hard to explain, but at least I'll try. So many things have happened for the better lately, even though at times seemed horrible, but then turned out great! That's life, it can't be planned or organized, person to person, or person by person. Shit happens everyday, planned shit too! So, you do the best you can with what you are given. Always stay positive and strong, regardless of your circumstances, cause tough times most definitely don't last, but tough people do. We all have opinions, and freedom of choice. Most people go along with the majority. Even if the choose is wrong, and has been proven wrong. Leadership is to do things different, before hand, not after the fact! My thing is that I will have the most patents of any African-American male in the history of the U.S. before I'm done.

I've got about 100 new ideas written down, but no cash to do anything with them right now. I need all of you to tell your family and friends that a truly blessed and gifted X gangster from West Eglestone wrote this book. From gangster, to best selling author,yeah, dreams do come true, and a lot of mine has, as well. These beings, I believe, along with God's hand, has kept me pretty well safe for life I've lived, so far. God has blessed me with the knowledge to pay attention to my surroundings. Maybe, I'm an alien hybrid, I I—AD/ don't know. I do know that I can remember a lady in white before my UFO experiences started in 1980..

THE END

CHAPTER 3

If it seems as if I'm going in a lot of directions, it's gonna come together in the end, o.k? U.F.O.'s may control human beings that been implanted from a great distance and I think they also watch us maybe as we watch reality T.V. Shows. Only for them we are the entertainment our life experiences. I believe this because I've seen some of the technology, they may be giving the government, at our expenses. The aliens must need us, or they are into torturing poor humans before they may kill us. Putting some people through horrific problems, bring some worst case scenarios true. At least that's what happened to me in my life. Even though I know it was going to happen, I could not stop all events I've envisioned. Heal s... too life the death of both my Dad and Granddad, Aunt Alice, and lots of uncles have died from everything from murder to pancreatic cancer of my best friend and uncle Mr. Kevin. When he died, I was in shock, did not drink or smoke. S.... deep Y'ALL and that's why I'm writing this book to make the world aware, yes aliens do exist and are here on earth night now and I can prove it. I also believe they can interact with us humans. I also believe they come a great distance to do what they do to us, for us, whatever they here. I've been investigating U.F.O.s for over 34 yrs. **Real Talk**, ever since my first remembered sighting.

In 1980, summer of Eglestone, I'm so glad that third phase of the moon, too the time to make a very big difference in the world of UFOs and alien contacts. The ET's seem to love some of the things we humans love as well like sex, and violence, what a coincidence, HUH? Well anyway, I'm gonna jump back and fourth and hopefully it will be published this way. It's my very first book of many to come, if we the people support the fruth. I don't believe in coincidences in life, everything happens for a reason, small, big, whenever and it all comes together in the right situation, hard to explain, but at least I'll try, so many things has happened for the better later, even though at times, I thought it was the worse, but some turned out to be great. That's life people, it can't be planned or organized person by person or for person for person. S... happens everyday. Unplanned s... too. So you do the best you can with what you are given or may already have. Always stay positive and strong regardless of your circumstances, cause tough times don't last y'all, but tough people do, Cause we all have opinions and freedom of choice, of us as a people to do as most, even if it's wrong and have been proven wrong.

So true leadership is to do things maybe a little different, before hand and not after the fact. My mentality is why I'm going to. Jesus willing, have the most patience of any black man on the planet, before I'm done. I've got 100 new ideas written down, but not enough cash yet to do anything about it right now. So I need y'all to tell your family and friends that a truly blessed X gang banger from West Eglestone, Saverly and Middletown wrote the book for the people from active front line enforcer to best selling author, Yeah, fam dreams do come true and a lot of minds has as well. These aliens I believe have interacted with me at times in my life changing fate at that time real BIZ. I've been through hell, high water, in sewers, put in hospital, jail, fights, shootouts, robbed three times, gun point of course, ain't no nigga taking from me without it way I thought and my team thought as well. Paying attention priceless and is the reason I'm able t o even share this info with the world. My third eye and the alien implant makes me the first alien hybrid, that knows who they are and must let the people know too. Am I the only one most likely not I don't know but I do know that I can remember a lady in white coming to my bedside at like 3 or 4 yrs old. My very first UFO experience started in 1980. Now when I was 3 or 4 and would run in my mom and dads room, screaming and crying, about the lady in white over my bed or crib a lot on Springdale Ave. 5600 (hundred) block I believe. So yeah maybe, I mean I was and Still am an out of this world reader. My math skills were not that great at first, but yes even they improved with time. Fact, I'm strong enough to do 500 push ups straight on que . I'll do them straight without stopping. Cross legs, lock around and under 5 min. My eyesight is 20/10. This is almost perfect.

So if you are a skeptic and your eye sight is not that, you may have a hard time seeing the info. 20/20 was my vision before I captured the beings in my neighborhood. 18 months after a visit to my doc, I checked the eye chart at proper distances and got all of them right on the very next to last line on chart. 20/10 she said and I got upset because I thought at the time, 20/20 was the best.

Boy was I shocked when the lady said Mr. Truth 20/10 is better than 20/20. Then I calmed down . I knew this was real info here to help us evolve. If it improved my eye sight, when it was pretty good at 20/20. What may this info do for people who wore glasses and contacts like my wife and step kids, which I tell them all day cause I raised them for 16 yrs. Been with my wife 16 yrs. Now after reading the alien info for over 2 yrs. Still finding new beings and crafts in the photos. So there's good ones and bad ones and I think they are at war and have been for quite some time, I believe. I mean, I think these alien beings, some of them, love me. Some of them are protectors of me and my family. But no one in my family, other than Kevin, would have seen what I am being shown, other than me, and maybe Fifgy. But my point is based on how I knew, I knew they were real, and let me know it. It's crazy and I can't explain it. It's like capability, I mean I have been observing them observe me. Now how long does it take us to receive info of our selves.

Photo's we look, we like, we don't like, we move on. But real aliens info, ain't that simple. There's 1000 of hours of research I put into this. Still to this day, I swear to God, Jesus, and many here, info still show's up.

So this is a on going process still taking place right now. So I hope y'all are really really ready for the truth people cause it's we are not alone, flat out. Ten or no less than ten alien species are here and I have them on film for all to see. No I'm not perfect, I've made lots of mistakes, but the one we cannot afford for me to make is to miss this info, for this generation like movie [knowing] o.k.

Nicolas Cage. Now from what I'm seeing y'all, there are good ones and there are bad ones! Aliens that is, just like humans, o.k. This is going to sound weird, but true, o.k. They seem to be at war for planet earth end of days type battles I see. Some of them are here because they know alien intentions of Dragonians, Greys, Marte, Mantis beings, Paladians, Nordiacs, they are all here people. They live in my hood for sure. They are there but you need to know how to go about finding them. I should have picked up on this info years ago. I'm sorry to us all if I'm the only one with and waited too long to wake up. Galactic gangster have come to protect us and make sure the world get's this info to the people. There are very few in my family that even pay attention to info concerning UFO's and alien beings. Maybe Kevo, we have both seen UFO'S together and my little cuzz 50, sees them to this day. And they are and were caught up in earthly duties used to keep us dumb founded from forbidden knowledge. True knowledge of ourselves and place amongst the stars. Think, these beings are our ancesters of a great blood line of beings DNA not fully human our family guys coming to holla, o.k? Protectors of humans cause without the whites and the blues, I call them the colonization by aliens, which would have killed us all, probably. I'm here to help, as they are. But us as humans, has been so programmed to think that appearance is enough it's not how you can be fooled by what you see. LOOKS can be deceiving, people, but actions can be deadly. Since I was eight, I wondered about my first experience. I knew my life has not been the same since. I knew it's crazy can't be explained but I just knew certain things most of us would not even think about observing them, observing me. When I first realized what I was very very blessed enough to capture on film in broad light.

After seeing them off and on throughout life, the biggest UFO story of my life put me on point and changed me forever. Back on the block, my mob will tell I'm one of most honest niggas or person you'll ever meet in the street. So after proving to them I can see them, they chose me to give to the people of earth alien information, real alien info not CSI or tampered with material all real 2000% authentic like me. However this book is 95% non-fiction, 5% fiction cause we all need insurance. We the people shall view my info and draw our own conclusion, without persuasion from powers that be. What I have on film, is going to change history for our past, present, and future which has been told wrong to the people. This evidence proves that there are several alien races here on earth right now, and I can prove it. God and Jesus willing, I'm going to get more, because aliens know they say it to the right one. With the protection of the holy cross inside my soul, I will expose these beings both good and bad. I believe they are maybe after things, o.k alien dead spirits, human souls after death, dimensional beings, or all of the above. They are here, they are not human. I know that these beings can enter our world and interact with us using technology noway we have yet. I think they feed off of us, our energy, all of it, it's why they are here. We sent out a signal back in the sixties I think, and yes, it was answered. When they wanna eat, they create situations to eat from, just like us, UFO's. I've captured always 100% of time, when I see an alien or aliens in these crafts that are on the ground right now. I think any damn item they can use to help them here on earth. Also the sun seems to be a part of them being here as well. These beings are doing doing everything from fuck in broad light, to killing each other, just as we may do. Got aliens jousting before and alien crowd. Yes, one dies on film people. They do die! When I was standing outside my apt. On May 19, 2012, my daughter was moving back in with me and my old lady, I'm fuck hot pissed. Because we could not afford to put her in her own place. So me and my wife could be grown and not bother so much, Plus we had been beefing off and on for about a year cause I wasn't working, but had unemployment coming in. So I was not bringing to the table what I had been up until that time. My point is everything I'm down some one help me get back on point. My wife works for comcast, was making $100,000, yearly. I was at about $3,000 grand from odds ins and unemployment combined. She held s... down, but the time off I had allowed me to view the photo's I had taken, seen what I saw and kept taking more and more recognizing more and evidence, real alien's evidence show up our film, s..'s cramy, cause I had UFO's and aliens on film with some camera phone from 2010. But I truly did not realize that the aliens were on there until 2012. You see, I'm not to technical of a guy, I like simple shit I did not even know how to zoom in with a cell phone first. But after seeing, one, two, but three UFO's, in the same day, less than an hour, I learned. May 19, 2012, changed it all in my life, on that day, in less than 40 min. and capturing two UFO's out the three, I learned almost all I could do with cell phone quick. Oh yeah, if anyone from Sprint reads this, your phone, the one style that was similar to Black Berry, at that time was the cell phone that brought this info to me, now I bring it to the world. Helped me spot the unreal but very real images that's gonna turn out to be priceless. I mean, if a photo of J.F.K. sitting on a plane during his term can be auctions off for millions. Now, stop think for a minute! If I could provide real photos of real aliens, having sex, with a female alien on earth, that the world could clearly see, people what price would you pay?

When government says they don't exist. I'm just asking cause I have it real biz. Broad daylight, the male is bending over a female and looks like she love's it. I've got the first alien porno flick on earth, you heard me. Period. Yes, it's that but to the point. A purple UFO shows up as well, same photo. You can clearly see the females alien's face, loving every thrust from the big burly green body alien, hitting her from the back. It's crazy, but even crazier things have turned out to be very frue. Real evidence I have. I think that we may be very important for the survival of all races. We may be part of the galactic food chain and cosmic survival chain as well, if not, than why would the what seems to be good aliens come to help us. We must survive, so that they continue to survive, at least it's my opinion, since day one on earth. I mean they are posted up in areas that humans don't spend a lot of time in, so, skeptics, if we are not there all the time, than how do we really know they aren't there or don't exist. We don't, but I know they do. I've seen them, but Jesus, I've never seen or God, but believe in them both, and with good reason, could be so crazy that all three exist. These beings I can see and most skeptic, say they don't exist. Even though real scientist say that E.T. Has to exist, so if we respect and honor these same scientist and deemed them credible, keep it the same when it come to UFO's as well.

CHAPTER 4

They say, it's almost impossible to say, they don't exist, with all the planets that have been discovered lately, but now we have to look no further than earth to find real alien beings and UFO's, real biz. Now I have the proof they say has been eluding mankind knowledge and acceptance. My real evidence is 100% authentic and priceless. But for a price, I offer any skeptic, to study or professionally analyze it for a small amount of time. I mean the best this planet has to offer, I challenge them. We can talk cash offer, I will cash my first check for this, look Jesus willing. People don't have a clue and that's where I come in, like the sun, to shine light on the truth. Been blessed and shining real life on the truth, since early in my youth. But I'm gonna get back to the spot I'm claiming, you see the West Eglestone Story, takes place in and out of town. My experiences, some of them, I have yet to mention. But I will, so I don't know how these beings, yet affect us in our lives. I just feel like they do and it's in good ways and bad. I mean I've seen alien beings fighting, having sex, smiling, shooting other aliens, trying to interact with my dogs, I mean just crazy shit, mean if those people don't know bout this, than I don't know how they don't. You'll see plenty of undisputable proof. The good ones I believe, want this info out, but the bad ones don't want that. Same with us! I believe they need us to keep the balance straight of creation. In the glory! If these things are evil, than I think they are so intelligent, they turn us against us, for them, unknowing humans have no choice after a while but to obey, I think. Those of us they may seem uncontrollable, properly get implanted like me. Others, I think are killed in a explainable and unexplainable ways. If you guys have seen the predator. The first one, it hides before attacking, just like animals in the wild are here on earth. Life for us in the wild puts us at a disadvantage, while out of their natural element. What if earth really belongs to them, not us, and we are being fooled, as we do our animals until it's time to eat them? Right? Predators blended in the background of envisionment. We are fair game people and powers that know it and I think sacrifice us daily. Nothing on earth seems to be able to detect them, better yet to stop them . Cosmic psychology has been exposed but you got to know how to spot it.

TV programming is called that for a real reason, people. If they show us nonfictional

Technology and call it fiction, we dubbed, ALL OF US! They keep us separate to keep us from coming together as a unit and may be able to stop them. So we are not the enemy, Maybe God said love thy neighbor as you love yourself. Maybe that's why he or she knows we are not alone. Stay together, period. World war three is now, people, us against them, not us against us, ok?

WAKE UP PLEASE!

These aliens do mean real biz, people, DON'T BE FOOLED! Oh, they are here and too much technology, that I see to fight and win without unity. They do fight with other aliens and humans from what I see. They even kill each other on film showing me that yes they can kill and see, I got so much alien info y'all not going to believe it, but 95% of what y'all see and read is nonfiction. Ok? It's all too real, people, all of it. We are the prize and if you and yours is gonna have a chance to survive, we must first know they exist. So if you have been brain washed by those people, than don't feel bad or slow. They go out their way so we never know we are one people. Must unite is messages I get. I've seen them in my house, my car, my wife's car, at my job, shit even while frying to s..., they've showed up. Don't know how they affect us, but I think they do, and if this is the case, than those people that know they are here, let poor human's take the fall for real aliens crimes against humanity. Those people are gonna have to explain how a human can be blamed for a crime, if it's a small chance the crimes were committed not under their own mental power.

These beings I think have been taking my family for years. Starting with my mom back in 1954.

See, her and my God mom, at time, was on Evergreen playground one day, and she claimed that a big bright light appeared and she remembered nothing else. She says, her and my aunt or god mom, also saw other kid's too out there. The crazy thing is, the hospital I was born in is less than a mile from her sightings. Again, coincidence? I doubt it, people. All and everything happens for a reason. Not known right away, at the time, but eventually coming together at the end of the day. So don't be fooled, please. They are here, we need to be prepared, o.k? Once they come out and interact with us, if they do. My vision is 20/10, right now and my third eye has been open for years. I humbly say that cause without it, don't know if I would be here to share this info we all need to know as a race. These aliens must have some type of galactic love for me or something must have been the way I always lived my life with no fear of nothing, that may have entertained them, and they must have known somehow nothing would keep me from this info for mankind so it must be important. I believe I was chosen before birth and have been trained out of course of my life to receive the info I have on film. So, guess what team, I got it now y'all, can you see it for yourselves? This info may have helped my eyesight go from 20/20 to 20/10 in less than a year, just by viewing them. How is it that my eyesight improved with age? I'm glad that this happened and I did the experiment and it worked with me. No medicine and no doctors and my eyesight wasn't bad, just got better to my surprise. Meaning that some people may benefit from these photos from just viewing them.

Now, in one of my photos, for instance, I have an alien running from me out of view of my camera but I caught him and you can clearly see a mind device control system attached to it's head and it looks like a dude who has glasses on as he runs away. But what I don't understand is, how is it, I can't see them most of the tim without the cell phones camera? My third eye, must Ltd have been what allowed me to catch his ab in mid O.k. What I'm saying is these aliens its several different type of alien environment, so they can function on earth. 90% of the aliens I'm catching are very close to a UFœI believe they are use to travel in. So those folks who say they are fallen angels, well what spirit needs a vehicle to travel in. So it's my opinion they are not demons or devils, but they, some of them, look horrific to see in broad daylight. Others look more peaceful, but able to defend themselves if needed to a tee. So, yes. They do have alien weapons in their hands and make sure I see them on film. Some of these weapons you the people are going to t£ them for the first time ever. The real info alone should put my book for the people go to top seller's list in less than a year. But please know, it's not about the money, but it won't hurt, so I thank all who support me and this very important cause for our race. Honestly I can't say how much they affect us on earth or our reality, but I believe they most definitely do and its up to us as humans, to deal with it. But no one, no matter who they are, can fight off that which we can't see, but may always be there trying to accomplish any goal they set to be done by us, to us. So I thank Jesus for giving me 20/20 vision to spot them with and after about a year my vision increased to 20/10. That's standing 15 feet away and seeing every letter of next to the last line of the eye chart. Got one before the last line wrong, that's it, 20/10. That's no lie and I have that type of vision to this day. It's all about EVOLUTION at least for my eyesight. Now at what cost though? I don't honestly know yet. Still trying to figure out if they are good or bad or both. I've seen both. Some people say they are fallen angels that God cast out of heaven. If that's the case, than why would fallen angels need UFO's to travel. Fallen angels may here too, but I believe, these are bonafide real UFO's and aliens. Maybe they are from another dimension, don't know, but I do know they aren't human and that makes them the most interesting thing on earth we have known D today. I think they have been here for awhile, too. I guess, I just noticed them as I did until May 19, 2012. I don't understand while more people are not seeing UFO's and their alien occupants. They are there truly. I mean it's so amazing to know they exist and not have to wonder anymore. I mean to know they exist when so many say they can't exist makes this info even more valuable. I have real evidence proving these things are real and here and I got them 1000's of times on film. Committing crimes against other aliens so if they got the ability they will affect us in anyway, that helps them, otherwise, why are they here? Any human caught on film, doing or committing crimes, they go to jail. Hardcore film have convicted millions of humans so the same thing that imprisons you, shall set you free. Period. Meaning if aliens are here, and more intelligent than us,than are all humans making decisions on their own in everyday life. If we are convicted of alien crimes, than it's time to change the way the law is written. Law is written for human behavior, not alien behavior being covered up by humans that may knowŽat our expense. I mean before I started gang banging, on the streets of West Eglestone, I had UFO sightings starting at 8. Yet you still have people out there who have never seen one UFO in life just because you can't see something, don't mean its not there. Hard for me to imagine that) cause they come out on the regular for me. People who have done no research, are the main one saying they don't exist. You are lying, crazy, or stupid to believe this. Well I can now back all real UFO reports with my evidence and that's all we need. I don't know of many people that can say they got rich of real UFO info. I think that you must have everything written down or taped to prove this.in any court of law. So I've got that covered, so when haters say, it's not real, you wake them up with my info and watch the looks on their faces. People are so caught up with their human lives, they don't have a clue as to what reality even have a chance to be like. I feel sorry for them. A lot of the aliens I've seen, look very violent and may mean us harm. Yes, they have technology, we clearly don't yet, invincibilities, cloning, projections, hologram and mental devices and more some of which I can't even explain.

Yes, it's all true and you won't believe what you're, yes see,and you're mind reads. All real 100% True. All nothing fake about this info and I have more to show and tell y'all, true stories! Yes, I have an alien implant in my body. They keep track of me that way. I can remember when young being paralyzed seemed like once a night a lot when I was young. It got to me so bad, that I developed a technique eventually to break their hold. I still used it to this day. I can also make their craft visible when they don't want to be seen. I've spent 17 years on the streets of West Eglestone, doing me, getting money, girls, respect, and power. I've had it all and the whole time out there seeing UFO's through it all. Good and bad times. I've also found a way to control my premonitions as well. That was hard and I'm still learning, but I do realize, they show you the future. Now that's priceless if you get the hang of it. But I've made mistakes, not listening to them, Once I started paying attention, things worked out better from that perspective. This is my reward to my people.

CHAPTER 5

Y'all must have an open mind and heart to see them and then I think they determine if they wanna go further with you. This knowledge is alien, y'all, and its all real, too real to be ignored. I wonder if this info, I've discovered, has been discovered by anyone else this way. If so, why have we not been given a chance to see it and come to our own conclusions. If the bushes, the grass, the trees, the clouds, and the sky is all being used as pieces of paper for us to read off of, than that's very extremely interesting to all humans on earth. Right? Of course it is. It may mean that humans are a everyday reality TV show for alien beings or something very similar to such. I mean of these aliens or other worldly beings are not human, but some seem very human like. If they are here, than how did they get here? Who are they? Where do they come from? What does these mean for mankind? I can't answer all these questions for sure, but I can tell yall what I know so far. Now I can speculate, what or where do I start? How does this implant affect me and my life. I know when it comes to real life and death situations, my third eye works very well and helped me survive the streets so far of West Eglestone. It's my alien instincts which have been priceless to me, my team, and my family. Our lives may have been saved a few times in life cause my instincts acted before my mind. We all got money, earned our respect, and our power grew to the   k   ) point in Eglestone. Let's just say our unity had us respected by some of the most ruthless killers or accused killers, this city have ever seen. East and West. During the 80's and 90's, and even still today for the most part, did what we wanted. consequences of our actions. So, something was watching over us all as we did our thing and established our selves daily in Eglestone. I mean we are the best of the best of the under world or was. 0000G's backed us in the streets for life from us putting work in early and often. We had our share to keep up with bloody reputations all over Eglestone. East and West. Our unity, as a mob, at that time was unmatched the way we did it. If youfachwith one, that meant you with many. From the time I entered Catholic School until I graduated, we were made men. But we didn't know it at the time. Till this day, us that's still alive and free, is not many from my block, but our inner mob, the original four, thank God and Jesus, are still here. More so them than me and they know why. So, I guess we are still being protected by something that we can't see and maybe not hear. But, they are there as you all can now see.

We made a hell of a lot of money, and made a few enemies, lots of friends, and put in mad work as a unit and in diversities on the street as well. By the way, some of y'all may have heard of the block where I'm from. Saverly and Middletown over West Eglestone, right of Bloomingdale and Poplar Grove. We were the real gang, way before the culture was broad cast on TV. Check the murder rate from Edmondson Ave., Poplar Grove, up to North Middletown. If you were out there at anytime, you earned it. Buddy, nothing was being giving away except testers. O.k? Those from the streets know what testers are. For those that don't know, free dope and coke. Shtåíve were affiliated with real G's by the way, I was 8 or 9 in my hood. But had family members whom had been in before I was born. Uncles who had served jail time for crimes from murder of police to disturbing the peace. But the familiar gangs of the 70's and 80's were the Baker St. Gang, Punk Rockers, and Countless Gangsters in infamous neighborhoods throughout the city. No disrespect to Bloods and Crips, but back then, we thought these were real gang bangers too. Eglestone had none or known very little about them. But these days, everybody is a killer and a player. Eglestone was dangerous then. I had lots of UFO sightings even back then. My research leads me to believe that some humans do work for aliens on a daily basis on earth. I believe they can mentally control us and implant us as well to control us in some way, shape, form, or fashion. My implant was discussed by me back in 2012. Don't remember actually being abducted, but, I'm quite sure, it has 100 of times as I've seen UFO's, just as many sightings by others as well. I put my life on the line, gang banging for years. I did it and have no regrets. Except one, should have made a better effect to save my money and I did not. So, yeah, crime does pay then and for some people even now. If you had organization, unity, trust and real love for each other, which we did and still do, helped me make well over a million dollars out there, easily, as a matter of fact, we made so much money and I can't account for how much I spent over 17 years off and on of hand to hand sells out there. O.k? Jail here and there, of course that's reality of being a real hustler. 804083 to 99, I got money on our block. Yes, me and my man, Manny, Rashoo Alegrea. My brother for life, we teamed up and it was on forever. Mugs and loop. Us four were a team, and other comrades also joined us off our block and Normount Court Apts., we went all in. Raul Coeke first, than we revolutionized cocaine sells, after five years of setting that co we moved to the dope of coke sells, ready rock, y'all. It was over. So, my team, my mob, were the first niggas in our part of town to sell ready, period, that I know of in our town. So, we went by Bottom Boys, and niggas know not to come through if you ain't have a PASS period. So, we kept that mentality for life. We have never lost a beef in the streets, or a fight as a mob. Most dudes knew not to even think about trying us. But, this is Eglestone, so yeah, many tried and many failed. I humbly say, we were all blessed as a team the whole hood, if you were in. We all got mad money, that was our primary goals. Young, black, gifted, well respected and hell bf-oron a mission, if its on, we ain't plan on missing. So often getting money off cocaine for a few years, we established our ready clout in the middle eighties.

Rahoo, Mugs, loop, and me, we were the foundation of this taking place in our hood period.

Hamburger also was a key reason we got, and at that time. By the way, Lil Richy and Jazzy J, from Pressman, shouldLova as well! So as we went about our biz as men, at an early age, we blew up and hid from no one or nothin. We had so many customers, that we could sell out daily, real biz. That's $9000.00 worth of ready a day and all that money went to just us, for the first three yrs. of sales. $36,000.00 a day, between us easily made for first three yrs. We sold it. That is just off ready. We also sold coke as well. So as time went on, our home¥ were mostly selling raw coke and dope stopped, that some of them to sell the ready too. We had it all, nickels, dimes, twenties, on the rock. All day, 24/7, shop . Some customers would smoke it so fast, they were back every ten min for a while.so lets say I got ten bundles of nickels, $ 1000.00. But it's 15 ofus all out the same time and on some days, we all could sell out. We could line up next to each other, take turns on sales and everyone of us will sell out. We go shopping and the night crew come out and do the same for a while. We could roll about 3:00pm, they come out and all be done by 12:00 am. So at it's best time, we had at least $40,000 a day of ready sales coming through any given time, flat out. So we all stayed fly through High School. I was a fly a.. Lil dude my whole life from living I mean Jr. High, I had Gucci shoes, and a different pair of tennis of all types from Michael Jordans to light blue from Addidas, each pair costing over $100.00 each.I WAS BLESSED. 30,000 grand a day, easy in the hood, back in my hood, back in the 80's and straight through the nineties as well. Now that number dropped over the years. But I know even today, I know at least 10.000 daily possible still being made right now. So me and my team were trend setters in our hood, legends in the making. Now it was not easy at all, but we some have to maintain our positivg Raw coke sales had turned into millions of dollars, now on our block to be made forever. If you didn't get billed or put in jail for life. It was a very fine line, believe me. If you have not lived it, you would not understand . We lived it for sure. For example, true being, I was spending $90.00 a day on weed alone. 7 days a week that I can remember. O.k? $630.00 a week on herb. Now let's do the math. $75,400 spent a yr. On weed alone back then, but I can account for it. So my total profit is accountable, cause every time you ree up, you spend more to make more money, O.k? No less than the dollar up game, guaranteed if you can get rid of it. For example, you spend $225.00 on quarter ounce of raw coke, nite. ou bring back $600.00 when you bag it up. Almost triple 275% profit every time, you grab more.Mad money but not easy at all.

Let's say, you bring back that $600.00 twice a day, made that day 100% profit a day, tax free, if you could do it. Just offa quarter ounce. Better than Wall Street. O.k, cash yall daily. SS*some niggas could spend $800.00. Bring back 2,000 at will a day, at Saverly and Middletown was it and we had held up to tradition for our time on our block. We were the foundation might and muscle for ready rock being sold daily, even til this day. Who are we, well Rahoo, loop, Mugs, Hamburger, and me Du Du. First three years of it being around daily, only we profitted of it. A lot of people was afraid ofjail time, you got ifjust caught with very little. See we didn't cock it up every time we grabbed up. But if your caught with it or near it, your charged G. We were raised by 000G's. Dudes with mad street smarts. We watched, learned, and got a chance and took it and got paid in spades, trust me. S*was so pumped up, that we had 24 hour shop, dope, coke, and ready. One stop shop, and we treated people like we wanted to be treated and it paid off. You knew the early bird gets the worm right? We were out there like the postman, in all types of weather even taking my whole yr. Off from school when I was in the 1 l th grade. But was still smart enough to pass, even with being put out temporarily. It helped us establish our clout for life, yes, it paid off.

I can't lie, can't tell y'all how much some of our mob made, cause I retired and kept getting it. So I have always gone out of my way to stay near. So I believe we did work for these beings, just unknowingly. Maybe we still do from time to time. From the highest to lower level of people on the planet. So this book is a wake up call, to mankind, either we are gonna wake up, or we may not get a chance to later. Now here's a bet more, God knows these beings exist. The bible says, prepare for all things and unseen in his world. So I guess God has given me a Godly type of vision because I clearly see on film, the unseen forces the bible speaks of. It's hard enough trying to prepare for that which can see right. So imagine having to be ready for beings from another world or dimension.

CHAPTER 6

Most humans don't even now exist. But this book, show's some of these beings that definitely some to be there, clear as day. These beings are connected to UFO's as well. 87% of my photographic evidence shows this o.k. They are here but where do they affect our lives clearly. Just a few questions. I'm going to try to answer for our race by books end. Thanks Page Publishing for a real shot at accomplishing my life dream of being helpful to our people in some good way. Now it's time to focus on the big lost of human life. This is just my opinion, the government, I believe knows a lot more than they are telling the people. But no matter what they do or will admit, must be so horrifically, terrifying, that we the people, must not be told at any point, period. When in reality, we have been left in the know not, but yet we the people make the government. That truth is, we as humans, are really up against a more superior species. I mean, we can't be the big dogs on the block, if something can just come along and show up at any given time and show us. We are being preyed on daily by sum of these beings. Come on, if you take most of us out of our natural element, we won't survive. Like new born babies, who must depend on others to survive for awhile. How do we adapt, to that which may be unadaptable. Humans are great at reacting, what I mean is, they wait to see what happens first, then react to it, the situation accordingly. I'm very good at finding solutions to problems that may or may not even exist. Some now, I believe my experiences, with these beings, whom some of them have told me they are from Mars. An whom goes with me everywhere I go. I believe, to protect me, or maybe even feed of me or both. They are attracted to higher energy, people. Whatever the case may be, I'm glad I realized they are here, and some, I believe, love us as well. Our space ancestors, brothers and sisters, yes, I said brothers and sisters too. Y'all will see both male and female beings in this book. Some of the women I've seen, ain't bad at all, yeah, maybe if desperate. I'd ///// Spotting UFO's at night was easy because we shot out or they would already off because the city didn't replace the burned out lights on the regular.We would see starlight objects moving as if they were on a cosmic highway. One night, we had about thirty of them over head moving slowly, and we all saw them. So, yes, UFO's were in our hood as I became 14 on the streets of West Eglestone. My home did not believe in UFO's ass I do, I don't think. We have never discussed them other then we knew they existed.

As we got our clientele in order while skipping school, we also had to deal withsights, shoot-outs and home drama. We kept force while we had 24-7 shop and was bringing in hundreds of thousands a year, probably in total profits just off our block.You see, we were the first niggers to sell ready rock in West Eglestone period. That's the truth, so yeah, my G's that's our title. I don't believe any one in the city, even to this day, made that type of money on a 15 year run. Going to jail a few times along the way, really wasn't part of the plan. I would return home, start the process all over again, never skipping a beat, and we wanted for nothing, the city was ours!

CHAPTER 2

Still until this day we have strength in numbers! We all were good at what we did, all soldiers on the front line, but none like little Du Du, okay, and I took my street mentality to another level once I got serious about cash, we all did! We had old heads behind us, young guys, guys our age, and ones that could not wait to get paid. We were called the bottom boys, readdy boys, Saverly and Middletown crew, but the bottom line was we were all street level dealers who made money daily. Fuck the law, stick up boys, spending time with women, all that. Just cash and fast! I stayed on the block daily, for hours at a time, only leaving if I sold out, or got locked-up. Trust me on most days, I sold out at will, on the bundles of ready.

To give ya'll some type of perspective of the cash, by 1990 1 was living in B-More County at 3600 Bowers Ave., Apt. D. Now, back then I was paying about $900.00 a month for rent, gas & electric, cable, plus I had a $90.00 habit f buying

herb for myself. *And my girl.* My habit alone, and I can't even tell you how much that cost me in a year. I started smoking weed at the age of 14 and Roxbury were my cigarette of choice. So, lets say $90.00 a day for weed is $630.00 a week, so you can imagine I   well over a mil, so yeah, we were getting it for sure! I;m sure I spent at least $30,000 or more a year on urb alone, I'm not sure you think maybe it was the weed messing with my memory? So, I would say on any given day, once we all established clout on the block you could easily have 15 different color vals at $10,000 a piece and everybody sold out at our peak, the math speaks for itself, $10,000 times 15=$150Ï000 'àdày on our block at one time!

Real talk people, so you can see a million was made on our strip many times over.

Even before we started getting money, G's were out there before we were Hustling was a way of life, but so was violence and chaos. Now on our comeup (3.1.27 ꞷ aíå several other notorious hoods, our West was the most deadly around. Gun play, yes, sometimes, but back then, you can beat a nigger up and not worry sometimes. When gun play was involved, well, that was an entirely different scenario! Young and dumb, and of *cum* course, you know the rest. But, we had respect, cars, women, clothes, Tims, and jewelry, I had my own crib as well.

Now, we were close, very close, so we knew about each others goods, bads, everything at one time. There were several OG's whom my mob looked up to in the early days. These niggers seemed to have it all, and our backs as well. We were well connected and still are, but for the record, I'm retired, so I'm gonna keep it real! Bob, Allen, A.D., Lil Kirk, Cabo,Lil Remy, Lil Leg, Joey Brinkley, Mike Brinkley,

Lil Jeff, Lil Dana, Tank, Pookie Shaw, Lil Jimmy, Kevin P., Fat Larry, Dante, K.J., Adolf, AKA, Maddog, Hitler, Nut, Michael L, Lil Toby, Ham, Pokey, Nut, David H., Coakley, Lil Rodney, Troy, Flip, Pie, Tiny, Yvette, Tay, Mugs, loop, gat Derrick, Lil Ricky, Sammy, Duane, Eurkie, Lon, Black, Stenton, Cassandra, Donna, Tisha, Kee Kee, Michelle, Joy, Lil Joe, Lil Joe, Lil Mal, Gregg, J. Hines, Lil Nanny, Fat Paul, Cornelias, Harvey, Raon, Razor Rod, Stan Dog, Nick, Nat, Ms. Bawanda, Ms.Nicey, Aunt Teresa, Samy, Keith H, Dante, Black Eric, Fred, Moe, Poopie, Weenie, Von, Pig, Butchie, Bandit, Malik, J.T, Pop, Dirt, Rue, Uncle B, Uncle B, Uncle Punnell, Uncle David, okay, that's enough. All these peeps touched my life at one time or another and a lot of them are dead and gone, but I say may God bless them all!I love you all and this experience and if only I knew then, what I know now, maybe some of them would have survived. You know the streets and believe me, UFO's were seen during trying times in the hood, at least in   hood. I was not the only one whom had seen them!

So, with the money came bullshit, jail, bail, lawyers, probation, parole, judges, courts, and of course money. We knew the rules before we joined the business. Staying true was the good, but was hard to do for everyone, but somehow, I did it. Yes, I made mistakes, but you never disrespected a OG who came before you. You don't disrepect them because you don't know what they went through before your time. Right there, on the same block where you reside, you see money! Stay true to your crew, no matter what! Follow the rules and stick to them. Now UFO's seemed to show up in my life during difficult times. While I hustled to pay rent, and stay fresh, and hip and cool all round the table. It was important in my hood to represent well. I was lucky I had all that strength and honesty as well.

For instance, I've locked ass with a lot of dudes coming up mostly fighting, but a test of strength and I've lost in a real situation only once. So, I would say my lock ass record on this planet would be around 60-1. Now, real fighting I'm 150 pds. Know on the streets of West B-More. That's real life and destruction too! The lock ass lost, no punches were ever thrown and the big ass nigger went to the hospital and got 6 stitches, I in return, was out about 6 seconds as well. We were working hard and got into a verbal altercation that turned violent, quick. After I came to, I hugged him and he knew he had won! He now had to respect my G period and that's how I've lived my life. Respect first, and everything else falls in place.

When you haved it, and I earned it the hard way, and I loved every minute of it as well.

Some won't believe this story after reading it, but believe me its 95.5% true. If your wondering what happened in the lock ass contest, we were both trippin. I got tired of talking and walked down on him fašT1n about 3.9 astro seconds. I grabbed him, plus, he was about 5'8", 225 pds., I stood about 5'11? 153pds. I wanted to pick him up and slam him, but I couldn't.

His sweaty ass standing chest to chest, he says to me,"Come on shorty, you can't get me like that," so I grabbed a little tighter, slid down his legs, and in the process when I scooped him up in the air and slammed him down, I was on top of him, but my head went up under his shirt, and when my bare face brushed up against his dirty sweaty ˢᵗⁱⁿᵏʸ ᴬˢˢ chest, I almost passed out! He grabbed his shirt, with my head still inside, and held on for dear life. He stunk so bad, I passed out in 2 milli seconds. Thank God he didn't know I was out and couldn't take advantage of the situation. In the seconds I was out, he got away from me. When I looked up, my knife was in the dirt. I usually kept two knives on me after my conviction in 91. This incident took place in 07. When I slammed him to the ground a second time, a piece of metal caught him in his back, and thats how he came to need six stiches.

In 1999, facing three felonies at the same time, for the second time in my life. I went to jail. After my three year stay, I came home and never sold cocaine as of yet. Yeah I saw UFO's before and after that incident! I have never really lost a contact situation other to my right hand man Sam Cook. He was like a brother. Sam Cook was better than me in a number of ways. Now, please don't think that I'm going in a lot of directions, but I promise, its all going to come together, ok. UFO's may control a lot of human beings who have been implanted from a distance, and I think they watched us on the regular. Just as we watch reality shows, only for them its us! This is what I believe, the technology, I know we don't have yet on their level. If they are working with the governments small versions. I believe just what they are capable of. They must need us, they are into torturing poor human beings before killing us, putting some people through horrific problems making worse case scenerios of thought come true. At least, thats what happened to me.

Several times in my life, even though I knew it was going to happen, I could not stop the events, real shit too! The death of my father, Gary, and my grandfather, Albert, my Aunt Alease and lots of uncle's, died from everything from murder to pancriatic cancer. My best friend was my Uncle Kevin, he was the closest uncle to me. When he died of pancriatic cancer, no amount of smoke or drinkcould ease the pain of his loss. Kevin is the reason I'm writing this book. Shits deep y'all, be aware that there are beings out here that can touch our lives, and as a matter of fact, I believe the came from very far away to do so. I've been investigating UFO's for over 34 yrs. Real talk, ever since my first sighting in the summer of 1980.

I'm so glad that Third Phase of the moon took the time to make a very different in the world of UFO's and alien contacts. The seem to like many of the same things that we humans enjoy. Things like sex and violence, what a coincidence, huh? Well anyway, I'm going to jump back and forth and hopefully, it will be published that way. This will be my first book of many to come if people support the truth I don't believe in coincidences in life, everything happens for a reason, big, small, whatever, and it all comes together in the right situation. Hard to explain, but at least I'll try. So many things have happened for the better lately, even though at times seemed horrible, but then turned out great! That's life, it can't be planned or organized, person to person, or person by person. Shit happens everyday, planned shit too! So, you do the best you can with what you are given. Always stay positive and strong, regardless of your circumstances, cause tough times most definitely don't last, but tough people do. We all have opinions, and freedom of choice. Most people go along with the majority. Even if the choose is wrong, and has been proven wrong. Leadership is to do things different, before hand, not after the fact! My thing is that I will have the most patents of any African-American male in the history of the U.S. before I'm done.

I've got about 100 new ideas written down, but no cash to do anything with them right now. I need all of you to tell your family and friends that a truly blessed and gifted X gangster from West Eglestone wrote this book. From gangster, to best selling author,yeah, dreams do come true, and a lot of mine has, as well. These beings, I believe, along with God's hand, has kept me pretty well safe for life I've lived, so far. God has blessed me with the knowledge to pay attention to my surroundings. Maybe, I'm an alien hybrid, I I—AD/ don't know. I do know that I can remember a weird body in white before my UFO experiences started in 1980..

THE END

12

CHAPTER 3

If it seems as if I'm going in a lot of directions, it's gonna come together in the end, o.k? U.F.O.'s may control human beings that been implanted from a great distance and I think they also watch us maybe as we watch reality T.V. Shows. Only for them we are the entertainment our life experiences. I believe this because I've seen some of the technology, they may be giving the government, at our expenses. The aliens must need us, or they are into torturing poor humans before they may kill us. Putting some people through horrific problems, bring some worst case scenarios true. At least that's what happened to me in my life. Even though I know it was going to happen, I could not stop all events I've envisioned. Heal s... too life the death of both my Dad and Granddad, Aunt Alice, and lots of uncles have died from everything from murder to pancreatic cancer of my best friend and uncle Mr. Kevin. When he died, I was in shock, did not drink or smoke. S.... deep Y'ALL and that's why I'm writing this book to make the world aware, yes aliens do exist and are here on earth night now and I can prove it. I also believe they can interact with us humans. I also believe they come a great distance to do what they do to us, for us, whatever they here. I've been investigating U.F.O.s for over 34 yrs. Real Talk, ever since my first remembered sighting.

In 1980, summer of Eglestone County, I'm so glad that third phase of the moon, too the time to make a very big difference in the world of UFOs and alien contacts. The ET's seem to love some of the things we humans love as well like sex, and violence, what a coincidence, HUH? Well anyway, I'm gonna jump back and fourth and hopefully it will be published this way. It's my very first book of many to come, if we the people support the truth. I don't believe in coincidences in life, everything happens for a reason, small, big, whenever and it all comes together in the right situation, hard to explain, but at least I'll try, so many things has happened for the better later, even though at times, I thought it was the worse, but some turned out to be great. That's life people, it can't be planned or organized person by person or for person for person. S... happens everyday. Unplanned s... too. So you do the best you can with what you are given or may already have. Always stay positive and strong regardless of your circumstances, cause tough times don't last y'all, but tough people do, Cause we all have opinions and freedom of choice, of us as a people to do as most, even if it's wrong and have been proven wrong.

So true leadership is to do things maybe a little different, before hand and not after the fact. My mentality is why I'm going to. Jesus willing, have the most patience of any black man on the planet, before I'm done. I've got 100 new ideas written down, but not enough cash yet to do anything about it right now. So I need y'all to tell your family and friends that a truly blessed X gang banger from West Eglestone, Saverly and Middletown wrote the book for the people from active front line enforcer to best selling author, Yeah, fam dreams do come true and a lot of minds has as well. These aliens I believe have interacted with me at times in my life changing fate at that time real BIZ. I've been through hell, high water, in sewers, put in hospital, jail, fights, shootouts, robbed three times, gun point of course, ain't no nigga taking from me without it way I thought and my team thought as well. Paying attention priceless and is the reason I'm able t o even share this info with the world. My third eye and the alien implant makes me the first alien hybrid, that knows who they are and must let the people know too. Am I the only one most likely not I don't know but I do know that I can remember a lady in white coming to my bedside at like 3 or 4 yrs old. My very first UFO experience started in 1980. Now when I was 3 or 4 and would run in my mom and dads room, screaming and crying, about the lady in white over my bed or crib a lot on Springdale Ave. 5600 (hundred) block I believe. So yeah maybe, I mean I was and Still am an out of this world reader. My math skills were not that great at first, but yes even they improved with time. Fact, I'm strong enough to do 500 push ups straight on que . I'll do them straight without stopping. Cross legs, lock around and under 5 min. My eyesight is 20/10. This is almost perfect.

So if you are a skeptic and your eye sight is not that, you may have a hard time seeing the info. 20/20 was my vision before I captured the beings in my neighborhood. 18 months after a visit to my doc, I checked the eye chart at proper distances and got all of them right on the very next to last line on chart. 20/10 she said and I got upset because I thought at the time, 20/20 was the best.

Boy was I shocked when the lady said Mr. Truth 20/10 is better than 20/20. Then I calmed down . I knew this was real info here to help us evolve. If it improved my eye sight, when it was pretty good at 20/20. What may this info do for people who wore glasses and contacts like my wife and step kids, which I tell them all day cause I raised them for 16 yrs. Been with my wife 16 yrs. Now after reading the alien info for over 2 yrs. Still finding new beings and crafts in the photos. So there's good ones and bad ones and I think they are at war and have been for quite some time, I believe. I mean, I think these alien beings, some of them, love me.Some of them are protectors of me and my family. But no one in my family, other than Kevin, would have seen what I am being shown, other than me, and maybe Fifgy. But my point is based on how I knew, I knew they were real, and let me know it. It's crazy and I can't explain it. It's like capability, I mean I have been observing them observe me. Now how long does it take us to receive info of our selves.

Photo's we look, we like, we don't like, we move on. But real aliens info, ain't that simple. There's 1000 of hours of research I put into this. Still to this day, I swear to God, Jesus, and many here, info still show's up.

So this is a on going process still taking place right now. So I hope y'all are really really ready for the truth people cause it's we are not alone, flat out. Ten or no less than ten alien species are here and I have them on film for all to see. No

I'm not perfect, I've made lots of mistakes, but the one we cannot afford for me to make is to miss this info, for this generation like movie [knowing] o.k.

Nicolas Cage. Now from what I'm seeing y'all, there are good ones and there are bad ones! Aliens that is, just like humans, o.k. This is going to sound weird, but frue, o.k. They seem to be at war for planet earth end of days type battles I see. Some of them are here because they know alien intentions of Dragonians, Greys, Marte, Mantis beings, Paladians, Nordiacs, they are all here people. They live in my hood for sure. They are there but you need to know how to go about finding them. I should have picked up on this info years ago. I'm sorry to us all if I'm the only one. with and waited too long to wake up. Galactic gangster have come to protect us and make sure the world get's this info to the people. There are very few in my family that even pay attention to info concerning UFO's and alien beings. Maybe Kevo, we have both seen UFO'S together and my little cuzz 50, sees them to this day. And they are and were caught up in earthly duties used to keep us dumb founded from forbidden knowledge. True knowledge of ourselves and place amongst the stars. Think, these beings are our ancesters of a great blood line of beings DNA not fully human our family guys coming to holla, o.k? Protectors of humans cause without the whites and the blues, I call them the colonization by aliens, which would have killed us all, probably. I'm here to help, as they are. But us as humans, has been so programmed to think that appearance is enough it's not how you can be fooled by what you see. LOOKS can be deceiving, people, but actions can be deadly. Since I was eight, I wondered about my first experience. I knew my life has not been the same since. I knew it's crazy can't be explained but I just knew certain things most of us would not even think about observing them, observing me. When I first realized what I was very very blessed enough to capture on film in broad light.

After seeing them off and on throughout life, the biggest UFO story of my life put me on point and changed me forever. Back on the block, my mob will tell I'm one of most honest niggas or person you'll ever meet in the street. So after proving to them I can see them, they chose me to give to the people of earth alien information, real alien info not CSI or tampered with material all real 2000% authentic like me. However this book is 95% non-fiction, 5% fiction cause we all need insurance. We the people shall view my info and draw our own conclusion, without persuasion from powers that be. What I have on film, is going to change history for our past, present, and future which has been told wrong to the people. This evidence proves that there are several alien races here on earth right now, and I can prove it. God and Jesus willing, I'm going to get more, because aliens know they say it to the right one. With the protection of the holy cross inside my soul, I will expose these beings both good and bad. I believe they are maybe after things, o.k alien dead spirits, human souls after death, dimensional beings, or all of the above. They are here, they are not human. I know that these beings can enter our world and interact with us using technology noway we have yet. I think they feed off of us, our energy, all of it, it's why they are here. We sent out a signal back in the sixties I think, and yes, it was answered. When they wanna eat, they create situations to eat from, just like us, UFO's. I've captured always 100% of time, when I see an alien or aliens in these crafts that are on the ground right now. I think any damn item they can use to help them here on earth. Also the sum seems to be a part of them being here as well. These beings are doing doing everything from in broad light, to killing each other, just as we may do. Got aliens jousting before and alien crowd. Yes, one dies on film people. They do die! When I was standing outside my apt. On May 19, 2012, my daughter was moving back in with me and my old lady, I'm hot pissed. Because we could not afford to put her in her own place. So me and my wife could be grown and not bother so much, Plus we had been beefing off and on for about a year cause I wasn't working, but had unemployment coming in. So I was not bringing to the table what I had been up until that time. My point is everything I'm down some one help me get back on point. My wife works for comcast, was making $100,000, yearly. I was at about $3,000 grand from odds ins and unemployment combined. She held s... down, but the time offl had allowed me to view the photo's I had taken, seen what I saw and kept taking more and more recognizing more and evidence, real alien's evidence s)ow up our film, s..'s cramy, cause I had UFO's and aliens film with some camera phone from 2010. But I truly did not realize that the aliens were on there until 2012. You see, I'm not to technical of a guy, I like simple stii.H did not even know how to zoom in with a cell phone first. But after seeing, one, two, but three UFO's, in the same day, less than an hour, I learned. May 19, 2012, changed it all in my life, on that day, in less than 40 min. and capturing two UFO's out the three, I learned almost all I could do with cell phone quick. Oh yeah, if anyone from Sprint reads this, your phone, the one style that was similar to Black Berry, at that time was the cell phone that brought this info to me, now I bring it to the world. Helped me spot the unreal but very real images that's gonna turn out to be priceless. I mean, if a photo of J.F.K. sitting on a plane during his term can be auctions off for millions. Now, stop think for a minute! If I could provide real photos of real aliens, having sex, with a female alien on earth, that the world could clearly see, people what price would you pay?

When government says they don't exist. I'm just asking cause I have it real biz. Broad daylight, the male is bending over a female and looks like she love's it. I've got the first alien porno flick on earth, you heard me. Period. Yes, it's that but to the point. A purple UFO shows up as well, same photo. You can clearly see the females alien's face, loving every thrust from the big burly green body alien, hitting her from the back. It's crazy, but even crazier things have turned out to be very frue. Real evidence I have. I think that we may be very important for the survival of all races. We may be part of the galactic food chain and cosmic survival chain as well, if not, than why would the what seems to be good aliens come to help us. We must survive, so that they continue to survive, at least it's my opinion, since day one on earth. I mean they are posted up in areas that humans don't spend a lot of time in, so, skeptics, if we are not there all the time, than how do we really know they aren't there or don't exist. We don't, but I know they do. I've seen them, but Jesus, I've never seen or God, but believe in them both, and with good reason, could be so crazy that all three exist. These beings I can see and most skeptic, say they don't exist. Even though real scientist say that E.T. Has to exist, so if we respect and honor these same scientist and deemed them credible, keep it the same when it come to UFO's as well.

CHAPTER 4

They say, it's almost impossible to say, they don't exist, with all the planets have been discovered lately, but now we have to look no further than earth to find real alien beings and UFO's, real biz. Now I have the proof they say has been eluding mankind knowledge and acceptance. My real evidence is 100% authentic and priceless. But for a price, I offer any skeptic, to study or professionally analyze it for a small amount of time. I mean the best this planet has to offer, I challenge them. We can talk cash offer, I will cash my first check for this look Jesus willing. People don't have a clue and that's where I come in, like the sun, to shine light on the truth. Been blessed and shining real life on the truth, since early in my youth. But I'm gonna get back to the spot I'm claiming, you see the West Eglestone Story, takes place in and out of town. My experiences, some of them, I have yet to mention. But I will, so I don't know how these beings, yet affect us in our lives. I just feel like they do and it's in good ways and bad. I mean I've seen alien beings fighting, having sex, smiling, shooting other aliens, trying to interact with my dogs, I mean just crazy sbbt:H mean if those people don't know bout this, than I don't know how they don't. You'll see plenty of undisputable proof. The good ones I believe, want this info out, but the bad åKn't want tÔÈSame with us! I believe they need us to keep the balance straight of creation. In the glory! If these things are evil, than I think they are so intelligent, they turn us against us, for them, unknowing humans have no choice after a while but to obey, I think. Those 1<11 of us they may seem uncontrollable, properly get implanted like me. Others, I think are in a explainable and unexplainable ways. If you guys have seen the predator. The first one, it hides before attacking, just like animals in the wild are here on earth. Life for us in the wild puts us at a disadvantage, while out of their natural element. What if earth really belongs to them, not us, and we are being fooled, as we do our animals until it's time to eat them? Right? Predators blended in the background of envisionment. We are fair game people and powers that know it and I think sacrifice us daily. Nothing on earth seems to be able to detect them, better yet to stop them . Cosmic psychology has been exposed but you got to know how to stop it. [i]

TV programming is called that for a real reason, people. If they show us nonfictional Technology and call it fiction, we dubbed, ALL OF US! They keep us separate to keep us from coming together as a unit and may be able to stop them. So we are not the enemy, Maybe God said love thy neighbor as you love yourself. Maybe that's why he or she knows we are not alone. Stay together, period. World war three is now, people, us against them, not us against us, ok?

WAKE UP PLEASE!

These aliens do mean real biz, people, DON'T BE FOOLED! Oh, they are here and too much technology, that I see to fight and win without unity. They do fight with other aliens and humans from what I see. They even kill each other on film showing me that yes they can kill and see, I got so much alien info y'all not going to believe it, but 95% of what y'all see and read is nonfiction. Ok? It's all too real, people, all of it. We are the prize and if you and yours is gonna have a chance to survive, we must first know they exist. So if you have been brain washed by those people, than don't feel bad or slow. They go out their way so we never know we are one people. Must unite is messages I get. I've seen them in my house, my car, my wife's car, at my job, shit even while frying to s..., they've showed up. Don't know how they affect us, but I think they do, and if this is the case, than those people that know they are here, let poor human's take the fall for real aliens crimes against humanity. Those people are gonna have to explain how a human can be blamed for a crime, if it's a small chance the crimes were committed not under their own mental power.

These beings I think have been taking my family for years. Starting with my mom back in 1954.

See, her and my God mom, at time, was on Evergreen playground one day, and she claimed that a big bright light appeared and she remembered nothing else. She says, her and my aunt or god mom, also saw other kid's too out there. The crazy thing is, the hospital I was born in is less than a mile from her sightings. Again, coincidence? I doubt it, people. All and everything happens for a reason. Not known right away, at the time, but eventually coming together at the end of the day. So don't be fooled, please. They are here, we need to be prepared, o.k? Once they come out and interact with us, if they do. My vision is 20/10, right now and my third eye has been open for years. I humbly say that cause without it, don't know if I would be here to share this info we all need to know as a race. These aliens must have some type of galactic love for me or something must have been the way I always lived my life with no fear of nothing, that may have entertained them, and they must have known somehow nothing would keep me from this info for mankind so it must be important. I believe I was chosen before birth and have been trained out of course of my life to receive the info I have on film. So, guess what team, I got it now y'all, you can you see it for yourselves? This info may have helped my eyesight go from 20/20 to 20/10 in less than a year, just by viewing them. How is it that my eyesight improved with age? I'm glad that this happened and I did the experiment and it worked with me. No medicine and no doctors and my eyesight wasn't bad, just got better to my surprise. Meaning that some people may benefit from these photos from just viewing them.

Now, in one of my photos, for instance, I have an alien running from me out of view of my camera but I caught him and you can clearly see a mind device control system attached to it's head and it looks like a dude who has glasses on as he runs away. But what I don't understand is,how is it, I can't see them most of the tim without the cell phones camera? My third eye, must have been what allowed me to catch his ass in midrun O.k. What I'm saying is these aliens live its several different type of alien environment, so they can function on earth. 90% of the aliens I'm catching are very close to a UFOs, I believe they are use to travel in. So those folks who say they are fallen angels, well what spirit needs a vehicle to travel in. So it's my opinion they are not demons or devils, but they, some of them, look horrific to see in broad daylight. Others look more peaceful, but able to defend themselves if needed to a tee. So, yes. They do have alien weapons in their hands and make sure I see them on film. Some of these weapons you the people are going to see them for the first time ever. The real info, alone should put my book for the people go to top seller's list in less than a year. But please know, it's not about the money, but it won't hurt, so I thank all who support me and this very important cause for our race. Honestly I can't say how much they affect us on earth or our reality, but I believe they most definitely do and its up to us as humans, to deal with it. But no one, no matter who they are, can fight off that which we can't see, but may always be there trying to accomplish any goal they set to be done by us, to us. So I thank Jesus for giving me 20/20 vision to spot them with and after about a year my vision increased to 20/10. That's standing 15 feet away and seeing every letter of next to the last line of the eye chart. Got one before the last line wrong, that's it, 20/10. That's no lie and I have that type of vision to this day. It's all about EVOLUTION at least for my eyesight. Now at what cost though? I don't honestly know yet. Still trying to figure out if they are good or bad or both. I've seen both. Some people say they are fallen angels that God cast out of heaven. If that's the case, than why would fallen angels need UFO's to travel. Fallen angels may be here too, but I believe, these are bonafide real UFO's and aliens. Maybe they are from another dimension, don't know, but I do know they aren't human and that makes them the most interesting thing on earth we have known of today. I think they have been here for awhile, too. I guess, I just noticed them, as I did until May 19, 2012. I don't understand while more people are not seeing UFO's and their alien occupants. They are there truly. I mean it's so amazing to know they exist and not have to wonder anymore. I mean to know they exist when so many say they can't exist makes this info even more valuable. I have real evidence proving these things are real and here and I got them 1000's of times on film. Committing crimes against other aliens so if they got the ability they will affect us in anyway, that helps them, otherwise, why are they here? Any human caught on film, doing or committing crimes, they go to jail. Hardcore film have convicted millions of humans so the same thing that imprisons you, shall set you free. Period. Meaning if aliens are here, and more intelligent than us,than are all humans making decisions on their own in everyday life. If we are convicted of alien crimes, than it's time to change the way the law is written. Law is written for human behavior, not alien behavior being covered up by humans that may knowŽat our expense. I mean before I started gang banging, on the streets of West Eglestone, I had UFO sightings starting at 8. Yet you still have people out there who have never seen one UFO in life just because you can't see something, don't mean its not there. Hard for me to imagine that) cause they come out on the regular for me.

People who have done no research, are the main one saying they don't exist. You are lying, crazy, or stupid to believe this. Well I can now back all real UFO reports with my evidence and that's all we need. I don't know of many people that can say they got rich of real UFO info. I think that you must have everything written down or even taped to prove this, in any court of law. So I've got that covered, so when haters say, it's not real, you wake them up with my info and watch the looks on their faces. People are so caught up with their human lives, they don't have a clue as to what reality even have a chance to be like. I feel sorry for them. A lot of the aliens I've seen, look very violent and may mean us harm. Yes, they have technology, we clearly don't yet, invincibilities, cloning, projections, hologram and mental devices and more some of which I can't even explain.

Yes, it's all true and you won't believe what you're, yes see,and you're mind reads. All real 100% True. All real nothing fake about this info and I have more to show and tell y'all, true stories! Yes, I have an alien implant in my body. They keep track of me that way. I can remember when young being paralyzed seemed like once a night a lot when I was young. It got to me so bad, that I developed a technique eventually to break their hold. I still used it to this day. I can also make their craft visible when they don't want to be seen. I've spent 17 years on the streets of West Eglestone, doing me, getting money, girls, respect, and power. I've had it all and the whole time out there seeing UFO's through it all. Good and bad times. I've also found a way to control my premonitions as well. That was hard and I'm still learning, but I do realize, they show you the future. Now that's priceless if you get the hang of it. But I've made mistakes, not listening to them, Once I started paying attention, things worked out better from that perspective. This is my reward to my people.

CHAPTER 5

Y'all must have an open mind and heart to see them and then I think they determine if they wanna go further with you. This knowledge is alien, y'all, and its all real, too real to be ignored. I wonder if this info, I've discovered, has been discovered by anyone else this way. If so, why have we not been given a chance to see it and come to our own conclusions. If the bushes, the grass,the trees, the clouds, and the sky is all being used as pieces of paper for us to read off of, than that's very extremely interesting to all humans on earth. Right? Of course it is. It may mean that humans are a everyday reality TV show for alien beings or something very similar to such. I mean some of these aliens or other worldly beings are not human, but some seem very human like. If they are here, than how did they get here? Who are they? Where do they come from? What does these mean for mankind? I can't answer all these questions for sure, but I can tell yall what I know so far. Now I can speculate, what or where do I start? How does this implant affect me and my life. I know when it comes to real life and death situations, my third eye works very well and helped me survive the streets so far of West Eglestone. It's my alien instincts which have been priceless to me, my team, and my family. Our lives may have been saved a few times in life cause my instincts acted before my mind. We all got money, earned our respect, and our power grew to the high point in Eglestone. Let's just say our unity had us respected by some of the most ruthless killers or accused killers, this city have ever seen. East and West. During the 80's and 90's, and even still today for the most part, did what we wanted. consequences of our actions. So, something was watching over us all as we did our thing and established our selves daily in Eglestone. I mean we are the best of the best of the under world or was. 0000G's backed us in the streets for life from us putting work in early and often. We had our share to keep up with bloody reputations all over Eglestone. East and West. Our unity, as a mob, at that time was unmatched the way we did it. If you fuck with one, that meant you fuck with many. From the time I entered Catholic School until I graduated, we were made men. But we didn't know it at the time. Till this day, us that's still alive and free, is not many from my block, but our inner mob, the original four, thank God and Jesus, are still here. More so them than me and they know why. So, I guess we are still being protected by something that we can't see and maybe not hear. But, they are there as you all can now see.

We made a hell of a lot of money, and made a few enemies, lots of friends, and put in mad work as a unit and in diversities on the street as well. By the way, some of y'all may have heard of the block where I'm from. Saverly and Middletown over West Eglestone, right of Bloomingdale and Poplar Grove. We were the real gang, way before the culture was broad cast on TV. Check the murder rate from Edmondson Ave., Poplar Grove, up to North Middletown. If you were out there at anytime, you earned it. Buddy, nothing was being giving away except testers. O.k? Those from the streets know what testers are. For those that don't know, free dope and coke. Shtit we were affiliated with real G's by the way, I was 8 or 9 in my hood. But had family members whom had been in before I was born. Uncles who had served jail time for crimes from murder of police to disturbing the peace. But the familiar gangs of the 70's and 80's were the Baker St. Gang, Punk Rockers, and Countless Gangsters in infamous neighborhoods throughout the city. No disrespect to Bloods and Crips, but back then, we thought these were real gang bangers too. Eglestone had none or known very little about them. But these days, everybody is a killer and a player. Eglestone was dangerous then. I had lots of UFO sightings even back then. My research leads me to believe that some humans do work for aliens on a daily basis on earth. I believe they can mentally control us and implant us as well to control us in some way, shape, form, or fashion. My implant was discussed by me back in 2012. Don't remember actually being abducted, but, I'm quite sure, it has happened 100 of times as I've seen UFO's, just as many sightings by others as well. I put my life on the line, gang banging for years. I did it and have no regrets. Except one, should have made a better effect to save my money and I did not. So, yeah, crime does pay then and for some people even now. If you had organization, unity, trust and real love for each other, which we did and still do, helped me make well over a million dollars out there, easily, as a matter of fact, we made so much money and I can't account for how much I spent over 17 years off and on of hand to hand sells out there. O.k? Jail here and there, of course that's reality of being a real hustler. 80 to 83 to 99, I got money on our block. Yes, me and my man, Manny, Rashoo Alegrea. My brother for life, we teamed up and it was on forever. Mugs and loop. Us four were a team, and other comrades also joined us off our block and Normount Court Apts., we went all in. Raul Coke first, than we revolutionized cocaine sells, after five years of selling that raw coke we moved to the dope of coke sells, ready rock, y'all. It was over. So, my team, my mob, were the first niggas in our part of town to sell ready, period, that I know of in our town. So, we went by Bottom Boys, and niggas know not to come through if you ain't have a PASS period. So, we kept that mentality for life. We have never lost a beef in the streets, or a fight as a mob. Most dudes knew not to even think about trying us. But, this is Eglestone, so yeah, many tried and many failed. I humbly say, we were all blessed as a team the whole hood, if you were in. We all got mad money, that was our primary goals. Young, black, gifted, well respected and hell <u>bent on</u> a mission, if its on, we ain't plan on missing. So often getting money off cocaine for a few years, we established our ready clout in the middle eighties.

Rahoo, Mugs, loop, and me, we were the foundation of this taking place in our hood period. Hamburger also was a key reason we got in, and at that time. By the way, Lil Richy and Jazzy J, from Pressman, should <u>love</u> as well! So as we went about our biz as men, at an early age, we blew up and hid from no one or nothin. We had so many customers, that we could sell out daily, real biz. That's $9000.00 worth of ready a day and all that money went to just us, for the first three yrs. of sales. $36,000.00 a day, between us easily made for first three yrs. We sold it. That is just off ready. We also sold coke as well. So as time went on, our homies were mostly selling raw coke and dope stopped, that some of them to sell the ready too. We had it all, nickels, dimes, twenties, on the rock. All day, 24/7, shop . Some customers would smoke it so fast, they were back every ten min for a while.so lets say I got ten bundles of nickels, $ 1000.00. But it's 15 ofus all out the same time and on some days, we all could sell out. We could line up next to each other, take turns on sales and everyone of us will sell out. We go shopping and the night crew come out and do the same for a while. We could roll about 3:00pm, they come out and all be done by 12:00 am. So at it's best time, we had at least $40,000 a day of ready sales coming through any given time, flat out. So we all stayed fly through High School. I was a fly a.. Lil dude my whole life from living G. I mean Jr. High, I had Gucci shoes, and a different pair of tennis of all types from Michael Jordans to light blue forms from Addidas, each pair costing over $100.00 each.I WAS BLESSED. 30,000 grand a day, easy in the hood, back in my hood, back in the 80's and straight through the nineties as well. Now that number dropped over the years. But I know even today, I know at least 10.000 daily possible still being made right now. So me and my team were trend setters in our hood, legends in the making. Now it was not easy at all, but we_____some have to maintain our positives. Raw coke sales had turned into millions of dollars, now on our block to be made forever. If you didn't get billed or put in jail for life. It was a very fine line, believe me. If you have not lived it, you would not understand . We lived it for sure. For example, true being, I was spending $90.00 a day on weed alone. 7 days a week that I can remember. O.k? $630.00 a week on herb. Now let's do the math. $75,400 spent a yr. On weed alone back then, but I can account for it. So my total profit is accountable, cause every time you ree up, you spend more to make more money, O.k? No less than the dollar up game, guaranteed if you can get rid of it. For example, you spend $225.00 on quarter ounce of raw coke, rite. You bring back $600.00 when you bag it up. Almost triple 275% profit every time, you grab more.Mad money but not easy at all.

Let's say, you bring back that $600.00 twice a day, made that day $1,200.00 made that day 100% profit a day, tax free, if you could do it. Just offa quarter ounce. Better than Wall Street. O.k, cash yall daily. Shit some niggas could spend $800.00. Bring back 2,000 at will a day, at Saverly and Middletown was it and we had held up to tradition for our time on our block. We were the foundation might and muscle for ready rock being sold daily, even til this day. Who are we, well Rahoo, loop, Mugs, Hamburger, and me Du Du. First three years of it being around daily, only we profitted of it. A lot of people was afraid of jail time, you got if just caught with very little. See we didn't cock it up every time we grabbed up. But if your caught with it or near it, your charged G. We were raised by 000G's. Dudes with mad street smarts. We watched, learned, and got a chance and took it and got paid in spades, trust me. Shit was so pumped up, that we had 24 hour shop, dope, coke, and ready. One stop shop, and we treated people like we wanted to be treated and it paid off. You knew the early bird gets the worm right? We were out there like the postman, in all types of weather even taking my whole yr. Off from school when I was in the 11th grade. But was still smart enough to pass, even with being put out temporarily. It helped us establish our clout for life, yes, it paid off.

I can't lie, can't tell y'all how much some of our mob made, cause I retired then and kept getting it. So I have always gone out of my way to stay near. So I believe we did work for these beings, just unknowingly. Maybe we still do from time to time. From the highest to lower level of people on the planet. So this book is a wake up call, to mankind, either we are gonna wake up, or we may not get a chance to later. Now here's a bet more, God knows these beings exist. The bible says, prepare for all things seen and unseen in his world. So I guess God has given me a Godly type of vision because I clearly see on film, the unseen forces the bible speaks of. It's hard enough trying to prepare for that which can see right. So imagine having to be ready for beings from another world or dimension. That's here but, we can't see.

CHAPTER 6

Most humans don't even now exist. But this book, show's some of these beings that definitely seem to be there, clear as day. These beings are connected to UFO's as well. 87% of my photographic evidence shows this o.k. They are here but where do they affect our lives clearly. Just a few questions. I'm going to try to answer for our race by books end. Thanks Page Publishing for a real shot at accomplishing my life dream of being helpful to our people in some good way. Now it's time to focus on the big lost of human life. This is just my opinion, the government, I believe knows a lot more than they are telling the people. But no matter what they do or will admit, must be so horrifically, terrifying, that we the people, must not be told at any point, period. When in reality, we have been left in the know not, but yet we the people make the government. That truth is, we as humans, are really up against a more superior species. I mean, we can't be the big dogs on the block, if something can just come along and show up at any given time and show us. We are being preyed on daily by sum of these beings. Come on, if you take most of us out of our natural element, we won't survive. Like new born babies, who must depend on others to survive for awhile. How do we adapt, to that which may be unadaptable. Humans are great at reacting, what I mean is, they wait to see what happens first, then react to it, the situation accordingly. I'm very good at finding solutions to problems that may or may not even exist. Some now, I believe my experiences, with these beings, whom some of them have told me they are from Mars. An whom goes with me everywhere I go. I believe, to protect me, or maybe even feed of me or both. They are attracted to higher energy, people. Whatever the case may be, I'm glad I realized they are here, and some, I believe, love us as well. Our space ancestors, brothers and sisters, yes, I said brothers and sisters too. Y'all will see both male and female beings in this book. Some of the women I've seen, ain't bad at all, yeah, maybe if desperate. I'd

///

beat em like they were human. This info is disclosure, its not cheap and was hell to capture, than even harder to understand, let me tell y'all. That's why $33.33 a book, we believe is a very fair price to pay for the truth. To all those that going to say this is or may be a hoax, behold, do some research and then give your opinion later, o.k? I've seen things y'all wouldn't believe, and that's why I'm gonna show the world as well. Show and tell. Extraordinary a million copies claims, takes extraordinary evidence, we have it. Our goal is to sell a million copies in one yr. God willing. Our goal is to sell more than a million copies in one yr. Once I've accomplished this, first part of this mission will finance more info to be brought to the world.

They all who are human needs to know, if we want to survive. Not just for the cash, yes, that is good, but for the countless humans whom lives may be saved, simply by realizing. We are not alone for sure. To prove this, I'll sweep all neighborhoods with my technique of viewing them there and what if they are the evil ones and can mentally manipulate humans in you're hood. This makes your neighborhood aware of what may be hidden unseen dangers. Expose to the people the alien influence possibility. Once you know they are there, you have to come up with a solution to get rid of them, or live with them as we do now. Most people are into God, so go to church, five days a week, then just on Sunday and see if it makes a difference. If you found out aliens were living in a big tree behind you're house, church daily may be a good idea, as a precaution. That extra time with the Lord. You would nothing ungodly would stand a chance. So If they mean y'all harm, that may work. If that don't work, then more then my God help us, real talk. I say that, because, I've seen them killing each other and there seemed to be enemies as well. Some of the aliens look happy, as they fly around on earth, with ease. I've seen alien foot soldiers, standing guard in front of their homes, right on earth. Some are watching me, as I take the photos. Oh, and big shout out to Sprint. SCP, 2700 version, because it takes the very best UFO and alien info I've seen here on earth, period. I want to be labeled the first man in Eglestone, to bring this real info to the people of the world. The first UFO and alien finder in history. So yes, we look to make history in many ways. Our services will be for sale to anyone who believes aliens may be around. If we find them, the photos to my customer, and they keep that photo for proof. If they show up, they are feeding off you and your family. So why y'all pay to stay, they take you're energy away, like it or not. True family precious factors, love, honor, and respect. Same as street code, which some of us live by. Follow that code and you should not have to worry about being crossed on the street level. Life is tough enough when you hustle and don't cross people, so you can only imagine how much more tough it may be. if you do. Now if you add alien influence into human life on top of that in which we are ready to go through, well all bets are off. I've seen Bigfoot, or Goatman, Dogman, Alien greys, Draconians, Insectazoids, Black Human looking type of beings, Nordics, or pleadians, they look almost perfect as well, all on film and in color, people. This is disclosure for the people from my 35 yrs of research on this subject. Y' all can view these beings some of them at bottom of this page. Just incredible photographic evidence, real evidence, o.k? So some people say, people who are having these experiences are crazy, delusional, seeing things, drunk, high or just down right mistaken. People please don't be fooled again by a government, who loves to use psychology for control. Cosmic psychology is used through movies, games, TV, Media coverage and just flat out denial, when they know the truth. People please do me a favor, use your common sense please. OK? UFO have been reported for thousands of years and are still being reported, even today, They have been seen in our skies, on the ground, on and under water, in space, I mean what left people.

The question we all should be asking is, where have they not been seen or reported on earth.

Every State, City, County, Country. Everywhere. What does the skeptics have to say about that.

OK? I challenge any skeptics alive to challenge info, its open for debate at anytime agreed upon, for the right price. Same goes for any talk show or movie producers or publishers as well. This is real undisputed alien evidences, deemed interesting even from our own government. They even kept about 30 of my photos I sent to them and they would not send them back and Mufon did not stand up for me either. After the visit from the Pentagon in the summer of 2012. Mufon viewed just one of my photo's and contacted the pentagon immediately. We meet at my house, me and the lead Mufon Investigator, well call them french. OK, come to my home, looked at my photo's and asked me about my life and my real UFO experiences I've witnessed. He also video taped the areas where I had the greatest UFO sightings in my life. He kept me anonymous as the record will show, and I'll keep them covered as well. But I was shocked when they exposed they worked for the Pentagon as a second job. So I knew that he proclaimed and seemed to be that real deal. So Mufon knew what I had was real. Despite the act they put on for public in some cases. TP&B have always been interested in any UFO and alien info, but real info especially, if they think its real after doing their research on it as well. My info God blessed me with receiving, was nothing but truth, people. Y'all are waking up hopefully already very amazed by what's here in color and black and white. I have enough info for three books, and three movies, and it's all real, at least. 95% non- fiction and 5% fiction, period. I welcome all people who buy this book that you will believe or have a better understanding by time you're done this book, just keep an open mind and put not one limitation, on God's ability please. God has no beginning and no end. There are no such things as coincidences in life. OK? No way. Remember God makes no mistakes, period. If you don't believe that, check the bible. OK? It's not the fact that things might be hard for you and you're family but rather how do you learn and bounce up out of it. Are you going to use you're ability to make a difference or make shit worse. Tough times don't last, but tough people do and ain't going nowhere, even knowing a alien presence are really among us on a daily basis. For those who still don't believe, well I guess you are just gonna have to wait until it happens to you.

I believe it will because they need to control us, some hour and for some alien reason or reasons. This is what I believe, I found my alien implant, and yes, I believe it still works as needed. I don't know how I help them or they help me, but I'm glad I realize they are here. The truth really is out here and I've found it by simply paying attention to my surroundings. The third eye, we call it on the street. But mines seems to have always been hightened for reason during time of danger. You know that I feel like something is going to happen feeling. A warning, something like a dream or even at the moment of action. Evolution is going to take place here on earth, but not all of us are aware enough to evolve at the same time. What I mean is, if you put limitations ofyou 're dreams, than you put limits on that which is achievable. To evolve is to get more inclined with you're surroundings. If it's something there, that which we can't see is normal, only until you spot it. But if you look the right way, you may be able to see it all. If you do show a family member, teach them. Cameras picks up on things that the human eye can't see, so the use of cameras, kind of gives us x-ray vision in a sense. If it's on film, than it's there, period.

Now we can explain whats there or what it is showing up, or where it came from maybe, or maybe not, regardless, it's there. Now if I can prove that aliens are here first, then I have to prove that the government knew about it. Now we have to really determine how they may be affecting the human race. If they are here and affecting mankind, man does not know they even exist. Are we in control or they? Even if it's possible they do control us even 1% of the time, then we have a even bigger problem because we would only believe that we are in control some what when we may not be even close. If these beings could control us just by 1% of the worlds population, would they use it to destroy as they please? Yes, from my research, there are good or bad aliens on earth and in the cosmos, has been the ----for billions of years before earth was ever around.

Any form of life once alive, strives to survive. Once you are here, it's all about survival right. It starts in the womb, as twins fight for food, before even being born. Sometimes, they both don't make it, but the one that does, kill it's sibling just to survive before birth. Now imagine, a alien race, hell bent on survival as well, but more advanced that we are, in every way. What would they be willing to do, to survive here on earth. I mean, if they are here, I don't think all of them, have our best interest in mind and heart. Some must have though or they are here to wait it out for our world to wake up or not. If not, than maybe they don't survive without our help forcing aliens and humans to co-exist for us both to exist at the same damn time. With only a few humans knowing the truth, but all aliens are possibly aware of us at some point in our lives and theirs. I can prove in any court of reasonable doubt that I may be right. If any harm comes to me or any member of our team, ton's of real photographic evidence, will be released across the country at the same time and I mean shit That you would not believe is possible, but is very real and true, I can prove this.

CHAPTER 7

This is not about money, but fruth be told, this real UFO and alien info is priceless. So we had to put the right price on it cause all real. Its not for everyone and it should be. It's all about if you want to survive or not. If so, than tell a love one about this book. If you need me to come to your area, to see if they are inn your neighborhood, I will. But I do charge $100.00 per visit, with a half hour limit per person, per block. I'm the first real UFO and alien finder on earth and I do this as away of making my living, recognizing, they are there or not. By during this, I think gives all of us a chance to make it. You can't be prepared for what you can't see or know ever exist. Not to look may be difference between life and death as a race. Can we afford not to know what we are really up against as a race? What's bigger than our race, we fight and kill each other daily, is tough enough. But imagine a real alien race of beings that's been here only God knows how long are here and hates mankind as some of the aliens I have on film faces show. Aliens that want most of us dead and PTB, work with them daily to make this happen. You know, they always have a way explaining the unexplainable on the regular. Sometimes, we as a people, has to use simple common sense. OK? Just because something unexplainable occurs, and we can't find the answer right away, then we fry and come up with something that makes us real intelligent, before we say, we don't know. Fact is, humans know very little about a lot of things that been here before us. We may find an answer to most problems that's how we evolve as a people. Experiences, living, Letting time past, to show us the right way, and the right answers. The people are not dumb but have been treated as if we are by our own people. Tell us the truth so we can at least pray an extra couple times a day to prep us and our family safe, Maybe you know don't just deny us of knowing what is really going on. We all need to know aliens are real, and not by announcement is needed to know it. But if those in power did the right thing and told the truth, the world may become a much better place. God probably did not start with us or stopped after us. God can do things, we will never be able to understand. I would think that the more He created, more praise He would receive. So the more He creates, the more praise He gets. Now before I started seeing UFO's in my part of town on a regular basis. My very first real UFO sighting occurred in 1980-83, just to be sure. After that, things changed for me. I mean, I was young, but well on my way to adult hood fast. Now, when we moved from Madison Ave. to 3215 Baker St., s... Got live. Around that age, school was most important, going outside and meeting other kids and people was the normal. Until then, my family were my friends too. I don't know if they are gonna like me or not. But I'm hoping they do. Now my step Dad and Mom, were hard working people, o.k? I just felt like I wanted to do what ever I wanted at that age. You know the age where you start to feel that way. Can't explain it.

Now me and my brother, Dirt, were new to the neighborhood at that age. But it seemed like a pretty cool crowd in our new hood at first, o.k? I'm in the first grade at St. Cecelia's Catholic School off North Ave. And Bloomingdale. First two yrs. of Catholic School went, o.k? Now, I'm in the third grade and the Nuns start realizing that I have reading skills better than most. Well above most kids in my class. There was about six of us out of a class of about 45 kids, whom reading skills was unreal. Example was in the third grade reading on a 8th grade level, no bull s.... I mean, out of the top six, I knew or heard of no kids reading for the Nuns, even after school. They would, along with my aunts and grandma come over just to hear me read. Sister Mary Joann, Sister Rosa, Father Sterling, all had been to my home or heard me read outside of school time. Knowing this early, helped me get here today. Now my Math, not so much. But I would get the hang of Math as well later. My brother was in the seventh when I was in the 3rd I think, so my reading was on his level. I never asked him if he remembered aliens coming up. My brother and I was close, but we never discussed UFO's, I don't know why. I also was expelled from Cathlolic School in the third grade. I brought a 22 revolver to school, and my classmate, I think his name was Claude, found it in the locker room, in my coat pocket, and told on me. When the Sisters approached me about it, both my parents were waiting. They were pissed because all I got was good praise in school until then. I had to come up with a story about forgetting it was in my coat from the day before. Now the real reason I had it, was because a lot of mornings we had to fight before school, as the Edgewood Boys would walk through our yard. They hated us because we went to Catholic School, and it cost a lot of big money, big then and still now. On a higher level than public schools. So we felt a bit special, as we should have. Fights would break out before, during, and after school daily almost. So I got put out for a while, but they let me lock in thank God and meet a few friends for life there. We stood up to them and my gun gave me all the confidence I needed. So yes, we would fight and I had a gun. We were Catholic, but not punks. You know they say the closer you are to God, the more the devil tries you. At that age was my mind set. Trust me, this is one of the most interesting books you'll ever read.

This is the West Eglestone Story by the truth. This book is 95% non-fiction. Just like my life has been, all real. I've been well tested by the very best West Eglestone had to offer. West Eglestone has taught me things you could never learn in no College on this planet. Learning to survive started early for me. The streets were waiting and wanting a chance to snatch my little ay5right on up. On Baker St., I learned how to play football, baseball, basketball, boxing, hide-and go seek, tag, skully, and all the games we were taught at that age. Priceless lessons of life learned, all while having real UFO and alien experiences as well. If these beings wanted me gone, they had at least thirty chances to do it thus far. I'm only

41 . I've been shot at twice, beaten by the police, robbed three different times, almost drowned five different times, almost abducted by humans twice, been to jail countless times doing about 7 yrs total with all my jail house stints. I've had dozens of women, about 125 fist fights, been tried on the strength tip 1000 of times in my hood and out. I've been banked by six people or more twice, 15 to 4 once, 5 to one once as well, all to my enemies advantage and never once did I loose. Somehow, no real harm has ever came to me or my mob when we are beefing with whoever. Now, I'm not saying people I know have not been gotten too, they have lots of them. My inner mob are all blessed to still be alive, me included. I mean, I bought my first nine millimeter at 15 yrs. old and vest at fourteen. Wore later daily for a while. We were also the first and biggest ready rock dealers of our time. We had already made 100's of thousands selling coke. When we moved up to ready, we went big time. By this time, we lived on Normount Ave. and Saverly and Middletown St. is the block we got money on me 17 yrs. off and on flat out. No one had ready rock in Eglestone but us for at least 5 yrs. We sold it all daily, coke too for a while, but that was a different animal cause we made about $30,000 off dope sales off and on. Seeing UFO's off and on while coming up during this time. Yes, West Eglestone had lots of UFO sightings and lots of drama as well. We all now established on raw coke and ready rock tip and some of us, the older mob had the dopes up North and Middletown as well.

Everyone see our success as a team. Other hoods wants to sell ready rock too.

Ellicot Drive came out with it. Lil Shauntes, and Pookie and them Ellicot Drive Boys, Poplar Grove Boys, came out. Bloomingdale and Saverly was us. Normount Court Apts. was us, period.

Edmondson Ave. Boys came out, Lil Paris and them. Poplar Grove and Harlem, Jimmy and Mort and them. Riggs Ave. Boy's. All of them got it now on the regular and of course other drugs were still being sold as well. Niggas or dudes know normally that could be seen as disrespectful. Most of these dudes knew us and would never even do that and sleep comfortably, so we all came together and competed with all of them. No block did it better than us. I mean, we had G's for decades in place to learn from and try to get paid but not make the same mistakes our 000G's did. Those right under us, could not wait to come up, Lil Stanley and them. Normount Court, Saverly and Middletown, played a major role of ready rock being sold in our town, Eglestone, Md., flat out. We had one stop shop for years, 24/7. I know I made well over a million dollars easy. When I subtract my jail time 10 yr run. Total of 17 yrs. We all got money together on our block. If I was home and free, I was on the block. My job was on our block. Yes, I'm a real X gang Banger, real biz. Not a rapper, even though I can rap a bit too, my goal to be a best selling author as well. I thank God for a chance to tell this true story, the real deal, people. UFO's and aliens exist, o.k? My photo's will prove it. Confirmed by our own Pentagon as legit. We worked for these beings, I believe unknowingly, for years as a gang, for real. We had older O.G's that helped keep us safe as well. All didn't make it, but all they instilled in us, will live for eveß This is just a few whom I believe the aliens may have also gotten too, through human violantile instincts. Lil

Marvin, shot in the head during a robbery, Big Marvin from Saverly, hit in the head with a tire iron during a fight and the club, Michael Lane, beat a dude up so bad, the dude got knocked out but when he woke up he left, went home, strapped up, and came back and shot Mike and killed him. Cokely, ran off with some cash and coke. The people caught him up Ellamont Apts. and shot him to death. Rodney from Westmont, got shot because he slammed a dude car door too hard after the dude gave him a ride. Todd, shot five times in the head and face on Middletown one night. Troy, from Saverly, was found tied to a tree fter five days in the park naked. Died from gang green a week later. My road dog Roy, I had just gave him a burner day before he got killed up Liberty Heights, Fat Larry, found dead in his cell in a fetal position, my man Big Kenny, shot dead down CBS, Big Tenance, my man homie, our big homie, shot and killed just days after selling us a Tommy gun, 88 in straight clip. Big Allen, one of the biggest dope dealers, even on this side of town for his time. Toby, my man crashed in a stolen car and bumed to death while being chased by the law. That's it for the book. May God blessed them all. There's more, but that's for the next book, hopefully. Look people, peace is the only thing, I think that can contain some of these alien beings, working or on through us. They feed off our energy, sorrow, pain, laughter, hope, dreams, they use it to survive I believe. Planet Earth is a feeding ground for aliens, and we are they prey daily. If you are not prepared, you don 't stand a chance in hell or Mars. O.k?

CHAPTER 8

If you are prepared, you may still not have a shot at nothing impacted by these alien invaders. Now I also believe that without their help, here and there, I would have been long gone. I just thank God for my third eye because it works, if you know what to look for. I've seen those aliens do the impossible, I have it on film. Evidence that they exist and they are here, maybe even controlling us to some capacity. I have a real implant in my body. I don't know who or what put it there, but what about those who don't even know they may have one too. This is not a movie, dream, or game, at least not to me. I mean I've been pretty hard on those who had crossed the line with me or mines. I'm blessed to be in a position to even write a book, let alone, one that I hope makes a big difference in both believers or non-believers, as well. I bet if you don't have an open mind, then I ask you, how do you even;see into that which whom you were born to be. If you have dreams and goals, then do that which they say can't be done. I'm also working on my first patient as well, so please guys, if y'all like this book, buy two please, o.k? Just kidding, buy three. Naw, but yeah, you can accomplish anything you put your mind to. I mean, I already with a lot of help from my mob, made as I said at least, over a million dollars on the streets of West Eglestone, easy. It was not easy, almost costed me my life more times then I want to mention. I would not change nothing because if one thing had changed, then, would I ever be here now? I believe, there is no such thing as coincidence. Every little bitty tiny thing happens for a reason, people. Some explainable, some not so easy understand or explain. It's all called life. Who are we to believe we are in control of anything past living? We may just believe that we are right because we believe in God. He controls all that is and yet to be. Yes, I'm a an ex-Gang Banger, but I'm still looking to make a difference. Why would God put us through hell, just to give us heaven? Maybe that's what we must go through to get there, people, just maybe. Explain why it takes a tragedy, a lot of times, to praise our one and only God? Everyone should live to at least be fair to the One who allows us to live. If life is as precious as we say, then why don't we go all out to be whoever we choose to be? Most are afraid of the unknown, but I think if we realize that it can be somewhat known, we would find out. Meaning that, looks can be very deceiving, so please pay attention to that which we can't see. Another world exist all around us and I've found it. Dozens of aliens in Eglestone County for real. I mean there is no denying the evidence I've collected and by the way, they helped me get them on film, why I don't know, but they did. I believe they are not willing to wait any longer to be acknowledged. They have been following me lately that know of since May 19, 2012. It's time that we realize maybe, just maybe I'm right, and if so, what does it all mean? Who are they? Where are they from? What do they want with our retarded ass. Obviously something wrong with them. Maybe we are entertainment to them in real life, I don't know, but they are interested in us, for sure. For anyone who doesn't believe me, just read this book, and look at the real deal photo's of these beings in action. I have videos as well, but that's for the second book Mrs. Diane, hopefully. If y'all help my family and I, may God bless y'all. No one is addressing this subject like me and it's time, too. As for the one's who wanna get all this info low, like Facebook and P.T.B. It's over, the truth won't be denied anymore. We the people make the government, pay them and yet they keep the biggest secret of humans kind to themselves, a secret. We all have the truths now in color photos and black and white. Go to alienanonymouseinfo@ gmail.com to by a piece of history. Your own evidence of proof aliens are boldly here on earth. They don't lay low all the time. Does this mean they do already interact with humans, not knowing, or are there others out there like me, who are very much aware of their presence? Yes, I say they can do just this, when they enter your house. I want to thank my family for not getting me committed to the nutty farm, when I first realized they were here. Yeah, I was hyper as fuck because once I realized what I had on film, I went nutty, y'all. Would you have been able to handle the fruth? I mean, believe or not, on May 19, 2012, they made direct contact with me and I got it all on film. Yes, you will see what real aliens on earth look like. Real Talk. This book is the truth and nothing but, at least 95% of it anyway. So as the first book comes to a head, if you are skeptic or believer, I hope you got your answer. Yes, UFO's and aliens are here and they seem to show up during violent times in my hood. Are they behind horrific behavior, don't know, but yes, there's a chance they are. They seem to be made of some type of energy, at least some of them. I mean, it's very weird because, they are mixed looking species, like cow and flowers, you know experimental coloring. I believe because they cloned me in a photo I took of them. Now that's suppose to be impossible, right, but it's been done, I have it on film.

They seemed to be watching us for some reason very hard though some don't look friendly, at all, but some do. How do we take what we see as we see it, or is what we see that in which is a image or light, they want us to perceive them as i don't know. Looks can be very deceiving from humans, let alone an alien race. They have shown me so much in such a short period of time. I know I am evolving into a better person for the planet. I mean, if you don't have an open mind, I don't think they 'll even seek you out. Maybe better for you if you don't, but if they don't get you, then what human pays what price, for you to be special. They must like me for some reason. I've remembered seeing a lot of alien info in my neck of the country, but this will change the way most of us see ourselves and the world-but me. I need not wonder anymore, they are here, and I can only accept this. How about the rest of y'all? Do you believe the people in charge who are using cosmic psychology on the people or is it all too real, to even begin to comprehend. Each person has to decide how they feel about it to get an conclusion. Just because we might all have different conclusions, don't mean that they are all great.Right? I'm now gonna make an offer to all skeptics, to first do your research, then compare it to my real alien info, or as I like to call them, alien pictures, puzzles, photo's that prove beyond a reason of any doubt. Aliens are here and very real, face it, accept it, or don't. If a picture is worth a thousand words, then the evolutionary process, would make these photos priceless, period. The biggest mystery, thought, wonder, feeling and so on and so forth. They are here and waking us up for some reason. Maybe even influencing our behavior as humans come from earth. Whoever gave the definition for human, now long ago and did not change over time, as we have. I mean, if my eyesight increased from 20/20 to 20/10 since viewing these photos, will it work for you? If so, what does that mean for me? Yes, i want to be nominated for Noble Peace Prize. Made man, inventor, best selling author, hopefully. Not bad, I'm working on it. Has this info helped me survive everything from jail, to gang banging daily? I've been just lucky maybe.What's luck? Can you see it, feel it? Does it exist? I say there is no chance that coincidences exist. When everything else seems planned. Right? Other native worlds naturally exist. Right? Well is that naturally planned or not? I'll have more info for the world if this book catches on. Book II will be coming as well, God willing. Why does the P.T.B. Hide the truth from the people? I mean, they must know the true intentions of these, or at least, some of them. When you see the expressions of the faces of these beings, its like they are saying, oh f..., he sees us, boys get him. If they got me, I don't remember it and they returned me, so whatever. I mean amazing photos and videos that the government says of things that don't exist, well I beg to differ, what I have really discovered. I've seen, and if I'm going over stuff I said before, well bite me, cause this is the final chapter for now. I've seen Nordious, Neptillian, Greys, Insectamoids, Big Foot, Dogman, Sheep, Squatun, Annunuki, Humanoids, monster type beings so ugly, I can't even describe them, but yall know. You can see them and get your own priceless piece of human history. Proof, aliens are real and here on earth right now. I can't disclose how I viewed my photos to find them, but it was all natural for sure, all real, no tricks, no gimmicks, no tampering. If anyone wants me to come on their show, to tell more of my incredible, amazing story, my bills still have to be paid and I'm expecting a second grandchild, so my price to tell the fruth is $10.000 per show, per head. You can't put a price on my life, so that's the charge, as insurance.

Four generations of my family on both sides, also have this info in case, God forbid, something happens. I also have family that is with the the F.B.I., State Troopers, EM's, and in Congress who will also be on point. If anything happens to any of us on my team F.O.G. This should not be kept from the public so we all shall know how to adjust to it, if we can. They seem to like sex a lot too. I took photos of them spelling out the word sex and it even showed up on film. You can see that as well. This is real people, o.k? I've got aliens coming throughout a dimensional portal in brood daylight. Yes, you can see this in the book as well. I've got alien beings having sex while I watch on film, crazy s...., but all real. Now imagine how much someone would pay to watch real aliens having sex in brood daylight. Priceless, I have it all on film. S... some of the beings look like members of my mob too. I can't explain our connection but they are giving me this info to share to you, period. I am not for free, no way people. This is worth the money because it's not that which you can see we need to worry about, but the unseen reality we live in may be more horrific than the one we see. Horrifying. I mean these beings, I've seen shooting each other in the head just as humans do. They seem to be attached to the UFO's as well. At least, 90% of them are near a real UFO. I guess they must stay inside of a portal like object to survive on earth. Some I've seen without a UFO but that's rare. They may be the clones I've seen, holograms, alien mind control devices on alien beings still show up to this day. If your eyesight is good enough. I believe they _____ in ways that naturally covered from questions. Like earthquakes, tornadoes, floodings, avalanches, cancer, morgelilons, and other diseases that cripple mankind. No cure for diseases. I believe A.I.D.E.S stands for Aliens Invented Dis S...Period. All I have to say about that. Oh, and they have the cure, but populations control would be affected if they gave that info out. Plus, what they believe about a lot of killer diseases is being called in question, so don't believe everything you hear as the truth.

CHAPTER 9

Now, cancer and possible alien connection, yes. I believe that UFO's let off killer radiation to humans. Invisible and odorless radiation. Just think, they can kill us all in couple of decades and never fire one shot. We never know what happened or where the cancer came from. Well, they live 50 ft. From my house and yes, I have these UFO's on film and videos. No bulls I'm also offering my services to find them in you're neighborhood as well. If I find them, the photo is yours to prove to people, you have the evidence of their existence in your vicinity which they should not be. I charge $50.00 per block, $100.00 per home sweep. All evidence you keep a copy of. If I find them, you and your family, then someone has a lot of explaining to do. 5000 people experience some thing shows a pattern. A lot of photos and any other evidences, I can find as well. Imagine if we find them all up and down East Coast with the special techniques of photography. Then what? How do we protect ourselves from invisible beings out harm U.S. First we must realize they are there, then you move forward. Acting like they don't could be a deadly mistake. I've seen beings that look like good beings as well, so there still may be hope. Yes, I wanna be the first UFO and alien finder on earth, period. Yes, I know how to find them. Just look at some of the facial expressions on a face, I spotted in my hood. I've got beings with three fingers. I've seen the inside of the UFO's, I believe on film. I've seen them on their planet, there at work, the colors, I remember were a very violet purple and tan, for the art work. There were three metal stick like objects in what looked like ships control area. I believe in my area, they are set up side by side, in my woods living inside the UFO. as a house or something. Silently sitting outside of our normal visual light spectrum. Cell phone camera, exposes them to us, they don't like that. I even had one alien show up in my 1989 Cadillac Eldorado with me in the house. It showed up without my permission. You know, one night after all this started, I had a feeling of being watched, so I was on point anyway. I grabbed my cell phone, aimed it outside my house towards my caddy, mean too, rims 20's, grey and black, two door coupe. Now my sprint SCP 2700, was ready. I also had my lawn mower in the car, storing it cause my property manager would not let me keep it out in the open. I laid a white towel over my lawn mower which was on the passenger side of my car and I also noticed a being showed up on film and it was watching me. No bulls..., I almost s...... my pants. I grabbed a stick, ran outside and saw nothing, but you know what I got it on film people, see this book is nothing but the truth. I'm saying unbelievable s..., but i got photo's of it and dozens of videos to back it all up. That's kind of special people. Thousands of photos of these beings that should not exist, but they do. Any skeptic that can de bunk any of my real UFO and alien info, I offer a challenge to you. If you can prove my evidence is not the real deal, I'll give you 10% of my earnings from part II. If I'm blessed enough to have this one be successful first. If you can't _____ my evidence, you give me 100,000 cash at that time. To sprint, AT &T, verizon, my night time photos, have some type of new technology embedded in the _____. I believe the future of cell phones technology. Yes, I'm open to discuss cash for them. I'm a businessman second and an informer of alien info for the people first. This info is all alien, very special, y'all. I know for sure I'm right, I've interacted with these beings as well but that I'm going to save for part II. So, yeah, I now have to inform the world, but the hood First.

Now people I'm telling y'all, Eglestone, Md. Has been colonized by an alien species, so stop beefing and please come together if you don't believe me, check out the evidence for yourself.These aliens love energy, all types of energy, but they seem to like violence like us as well. What if they are getting us to be violent towards each other, you know gang banging,and innocent people being killed?

What if they can control us at will? Let us kill each other while they're laughing, getting stronger and staying hidden while human beings take the fall for alienated crimes against humanity. If the universe is 14 billion yrs old, and the earth is a mere 4.5 billion to 6 billion yrs old, how in the hell you expect life, not to happen over tremendous long periods of time. It's impossible, people! Life can't be stopped! It just happens, o.k? If these beings are here, and I'm the only one that noticed, and got them on film, then what would that make me? Evolutionary, means to receive info of in for the people. We are evolving everyday, now we know the truth, we are not alone on earth, better yet in the universe. Now please, these hoods I need to so love too. First and foremost Saverly & Middletown, Saverly & Bloomingdale, M.O.B. Niggers are behind this product. Affiliated with 276, Smiley, Phazon, before s... went down, Lil D, R.I.P., Mad Dog, Adolf Hitler, I love you 0000G, Rahoo, Mugs, Spanky, Waffles, Dike & Dike, ___, Ham, Pokey, Nutt, R.I.P. J Rock, love you O.G., Steton, Troy, Brian, Rolo, David Wallace, Von-du, Tisha, Yvette, Annette, Lynette, Nat, Pooh, Elaine, Loranie, Aunt Yvonne, Aunt Teresa, Aunt Sandy, Nina, I got some on cash tip for all y'all in a minute. Miss Johnson, you did a great job, _____ her son Denick. This is for us all family! Nina and Shelly, thanks for taking care of my daughter if they are. Both of y'all. Maybe we can get the test done and know for sure. Still time for us all to come together if possible. I thank my wife, my two twin daughters, I do know. A and R. Shout out to my lil brother Picallo, Tiny, Miss Neicy, Miss Sandra, Miss Dotty, Miss Fleming, Uncle Billy, Uncle Paul, Manny, Lil Ty, Amena, Kee Kee, Boyett, Ice, Tammy T, K.T.,Celestine, Benny, Lil Ricky, Lil Stephie, Lil Pizza, Bird, Big Jeff, Lump, Mr. Adullah Griffin, Dirt, Brian, Ru, Aunt Janice, Uncle Brian, David, Uncle Ronald, All the Robinsons,, Taylors, Coopers and Browns. North & Middletown, Big Earl Big Legg, JR, Pudding, Cabo, AD, Lieman, Joey Brinkley, Mike Brinkley, Nookie, Malik, Jammy Jeff, Rico, Lil Parnelly, Quinton, Big Irac, Ronald Jones, Pop, The Village, The Junction, Lil Mousey, Rooster, Cherry Hill, West Port, L.T.,M.H.,Pig Town, Park Heights, Chadwick, Woodlawn, C.B.S & A,

Skeeta, QJ, R.I.P. Deno, Leroy, R.I.P. Cleo, Lil R.I.P. Tyree, R.IP. Leroy, Hendricks, Fat Larry, Troy, Todd, Flip, all R.I.P. Brother's real y'all, over West. O.k, lets look at the facts. UFO's, have been seen everywhere people. How can they not exist, should be the question. Common sense tells us that history is written only to be timely changed in some cases when someone accomplishes that. We thought the earth was flat, but it's round, we had nine planets, right? Imagine how many more scientist we may have now, if we thought that then. How did we evolve apes, when apes are still apes, y'all. Something fantastical happened with us and we are it, right now. Why are they here? Why are they hiding if they mean no harm? Why do the government claim they do not exist? Why do so many things go unexplainable? How many unexplainable situations have to occur before we realize unexplainable should be rare, not normal. If it happens a lot, it's no coincidence, no its not, people. Please wake up before you don;t get a chance too. It gets no clearer than this, if we keep beefing with each other, all the soldiers on the front line fall, so who is going to protect the women and the kids. Even the best of the best meet someone better, eventually, life enjoy life, love yourself, if your down, change your situation, you have the power. I imagine alien haters with powers that can control you. I'd want to know about that. How do we stop them? Well I don't know. We just have to be prayed up! If that don't work, then I don't know. I do know they are here. They mean biz too. At least that may be what they want me to believe or even tell the world. I'm going to tell y'all just some of the events that are UFO and alien related, may link up to me starting I believe around 3 or 4. The lady in white shows up from time to time and looks down at me. Then when I was 8, 5612 Wesley Ave. During summer of 80-83, just to be sure, my lil cousin Eric, one of my best friends of the family, and me, at the time Tee. This I told to the world. On May 14, 2014, On the third phase of the moon, radio talk show, my aunt had a cookout and a lot of my family came to see the new house and Tee, Eric, and myself, saw the same bright object or a UFO after we all moved out into the alley of the house. Now my aunt's house was directly in the middle of the block and when we noticed the UFO, me and her both looked at each other and asked, almost simultaneously, "did you see that." I asked and then she said, yeah, and she asked me, and I said yes. Then the next thing I truly remember, was still looking at it, then we all were coming up from the front part of the house at the right bottom part of the block, near the woods. Even at 8 yrs old, I knew what I was seeing was not normal. I don't remember how we got to the front part of the block. As we reached the house, I noticed the sky, it looked different for some reason. It looked different because it was now later in the day, as well. I can't say for sure how much missing time went by, but I believe it was around 12:30pm when we came out into the alley. Now it seems as if it was 4:00pm. We all went in different directions after we got there. I don't know why our parents weren't looking for us or why they saw nothing.

Tee and I, also won a house through an essay contest of a radio show in 2002 as well. She did most of the work, but I added a few things and she won. It was a house over East Eglestone. Plus, I was with her same year and I saw a golden torpedo type UFO one day as well, as we arrived in the city. It occured around

4:00pm. We were in a car at a red light, when I looked up and saw the object. Another time in 2006, in brood daylight, in the summer, around 6:30pm, my Uncle Kevin, my very best friend, and myself. May God bless our family, cause we lost a legend, but anyway, we were in the Park Heights area, as we were finishing up some house work, on my Aunt Jennifer Smith's house, it was now time to do some roof work. I really don't like heights, but climbing a ladder about 30 ft. in the air, did not set well with me or Kevin. I have rode some of the meanest roller coasters and been on several airplanes in the past, we both had a bad feeling about this. So, once I got about halfway up, Kev said, "Yo, you don't look comfortable, come down. Let me do it." He had never had a problem with heights though. So I came down and he went up. I was truly thinking about seeing a UFO and I had been listening to Jay Z for hours. I had a white 94 Acura Vigor at the time. As I watched him get closer to the roof, I saw a star appear, in the atmosphere, ok? I immediately yelled to him, "Yo, do you see that star flying in the daytime." He stopped, looked up, and said, "yes sir, I see it." We were just discussing past UFO sightings. He and I both had, had them. But how did the aliens know? The star was there for about 30 seconds, then a greyish metallic disk shape UFO showed up first, in the middle, as the star disappeared. Then one to the left of the middle one, showed up, like just appeared already flying, the third to the right. They were in a fighting formation, period. I have seen planes do the same thing in the past. I asked him did if he saw them and he did. They were about 15,000 ft up flying silently in brood daylight up in Park Heights, no bull s... I was amazed, happy we both saw them together. He had told me after our sightings, that he and my cousin Lil Jessie, was riding together up Park Heights in the past, and they both saw alien monster type beings, lots of them, walking the streets and it wasn't Halloween. Another time like in the early 90's, me and my MOB, were all outside, one night, getting it in and all the lights were out around Normount Court Apts. We all had been drinking and smoking herb. Only we got money for all other drugs being sold. Anyway, we all was looking up at the sky and saw star like objects that I spotted flying as if they were cars driving on the streets. They were flying in straight lines, intelligently flying, and we were like, "what the f....are those," and it was like close to a hundred of them too. Another night, we were out and we noticed an erie greenish light, neon colored light and fog booming from a wooded area. It was like five of us, ready for anything, and I said, "let's go see what it is,"and you know what, we never went to check it out. It was that weird. Another time, same area, like in summer of 94, me and my road dog, Mugs, were outside around the apartments, I showed him a metallic UFO hovering low in broad daylight. The disk shaped object was at least 8000 ft. in the air. A couple of sales walked up and we greeted them but don't know where the UFO's went.

I would say, yes, they are really here watching us for sure. I can only speculate on their reasoning for being here. My mom and godmother, Anita and Vivian, back in 1954, had a UFO sighting that my mom remembers as well. They were living on Glenolden Ave. And were at the Evergreen playground, when she said she remember seeing a big bright light that just showed up over the playground in front of them and other kids as well. My dad's mom moved to the same house some years later, where my dad, uncles, aunts and I also grew up in. I remember a few weird things happening there as well, but no UFO sightings for me there ever. Ruby Brown, my baby, and grandma, whose my mom's mom, I never discussed UFO's with. She died of cancer in 2004.

My dad Gary, had given $250.00 to a crack head. She came back for more and he said no. They fought, he f... her up and she stabbed him in the back while he was asleep. He died in the hospital 7 days later. She was convicted and received 25 years in jail, thanks to my grandma. Now let me also add, she had her throat slashed while she was in jail. Her throat was slashed from ear to ear, survived, and later became a preacher. Which was better than me and my mob catching her on the streets. He worked everyday and enjoyed his time off. He had just called me a week before he got stabbed. Normally, I rarely would talk to him. We knew each other but not as much or as well as we should have. He bought us all a lot of insurance money but I used mine to bury him. He was a Vietnam vet who fought on the front line at one time for you and yours I miss him and really didn't discuss UFO's with him. I've also had at least 15 uncles from his side die of cancer.

I've had four family members murdered right here in Maryland. One of whom was murdered and raped, my Aunt Alice, by her own nephew. My uncle Lump who was accused of her death; from what I understand, was hit by a train down South later. Bobby T Jr. Was killed by a vicious blow to the head with a lead pipe over East. He was set up by a female companion. T.T.J. was shot in the face, and killed in 2012. They got his accused killer. I didn't know them well, but blood is blood, may God bless them.May God bless us all, because if these beings are real, and can control our actions, how would we know what is what? How do we stop something that's not suppose to exist. Prayer is the only thing that may help and if it does not, we're in a lifetime of trouble, people. We must start to realize, and trust me, I tried to follow all codes of banging at all times and keep others on point too, I put all types of work in maybe being influenced by alien beings. I had to start our own organization called, F.O.G., "Family Only Group." We were the first non-violent group from West Eglestone. Our focus was to get money and enlighten other family of what we are really against in life. No gang violence but peeps in place if needed. My first legal goal, if Mrs. Diane blesses me, to sell one million copies, A.S.A.P. We will be able to come to the table with some of the leaders of the worst labeled gangs and offer peace to all gangs and their enemies to be peaceful for life. Once we are established, I the truth, will make sure a certain percentage of my profit will be paid to certain gang leaders in Eglestone for peace in the city. Yes, we have been recruiting F.O.G. members of peace.

We have a few members in a few States. Our goal is to pay for peace State to State, eventually. Some proceeds will go to find cure for cancer as well. I've also lost like 20 members of Mom's side from cancer over the years as well. These UFO's may be killing us and our kids early. I'm also an original M.O.B. 000G and is affiliated with real gangsters, 276, Bloods, Crips, M&M, Hell 's Angels, DMI, and lots of street gangs here and a few places out of town. Look, we must stop killing each other on purpose because real niggas die too on both sides and for what, rep., Respect, money and girls. These beings, when the time is right, are going to come after our women, kids, and all the men. Real soldiers will be dead or in jail and will not be able to protect none of their family from harm. I don't know if it would make a difference, but ir just may. There may be a chance, maybe not, but we are going to have to find out. I'm not into telling grown folks what to do y'all but if you do some research, you;ll see for yourself. This may be just be an explanation for the unexplainable, even if we have not realized it yet, doesn't mean it has no answer, my nose would start bleeding so bad, sometimes I had to leave school early more than once. I can remember some unseen paralizing force coming over my whole body on the regular at night. I mean thousands of times, I clearly remember. Also felt as if my breathing slowed while this was happening. It would happen so often, I got tired of it and developed a technique to counter the attacks. What I learned to do was focus all my energy on one major part of my body to break the paralymotion affect. But for years I tried until finally It broke my paralymotion. I don't know why or how but it works. These beings have the power to show up at will, if the light can be used for them to manifest themselves. I've got it all on film, the real proof they are here. I also explained on May 14, 2014, to the world on third phase of the moon. I believe I have an alien implant in my body. For one, when I was just starting to have sex, the women loved my sex drive. Yeah, they thought I was one of the best that ever did it. Now really because all the virgins, o.k, most in my hood around my age. They heard That I fvçh Well. So I was a first time guy for a few. A lot of women just wanted to fuck me and I turned down very few pussy. Being offered. Even if I had a girl. My having one women was o.k, but the more money, women, and action you got, the more respect was given. I was gettin money, at an early age, old money. Raw coke money, people. Now of course, me and my manny, Rahoo, were partners with Kevin, Ricky, real Og. We started out on salary and learned fast, holding coke, dope, and guns could be profitable while we waited for Rosemount Rec to open for fun. We were on our way to making over $1,000,000 on the streets of West Eglestone. We were MOB, before it was official. The original MOB were the first niggas to sell ready rock in Eglestone, period. Saverly, Middletown, and Penn Ave., was what I know of. We made so much money because everyone knows how <u>coke effect sold</u> out, its still going on today, in all hoods, ok. You know when niggas saw the cash, we were making, they came out with their own ready and before we knew it, we had competition from everyone, everywhere, but Saverly & Middletown revolutionized cocaine selling in Eglestone Maryland.

Our shit Was coming from a smooth usher face, Michael Jackson type dressing dude name Chris, who was from up top. He would catch the train with the real good shit on him and give it my manny, my nigga, my homie for life and one of the first dudes to trust me to keep that cash straight, Hammy. One of our first Bosses in our hood and in game, but on the low. He's with the Feds now, but will be home soon. He ain't no whore either. My road dog Rahoo and me, stayed with him until we earned our stripes and re-entered off on our own, like two years later. We made good money with him especially being young but we wanted more. We teamed up, Rahoo and me which was after we had sold raw coke for years well, and we teamed up and took all work, we could together. Mugs and loop, Spanky, Nat, Won, Steton, we all got it in together. Unity, you f...with one, you f... with us all and there's more of us that I won't even mention. Maybe a few legends, Adolf Hitler, Lil Eager, Chris H, my mentor and nigga for ever, Poky, Nutt, Heize, Big Black, Danny, his lil brothers, Mike Lane, Jealous Kev, Flip, Shermon, Fat Calvin, Lil Joe, Erikie, Harvey, Lil Pie, David Sudd, Rolo, David Wallace, and the rest of the young and older MOB mentors. Most of us stayed true down for life. Back in the day, as we banged, we were called the Bottom Boys. Don't let us catch that a.. in our hood and you know no one. Not good for you or whoever you wit. We kept and still keeps it gangster. These beings have made me put aside all beefs I had with humans but one, loop, I'm in that a.., when I catch you my G. Put the gloves on cause we fighting and you know why my nigga. Still love you though, yeah, but the knockout game we gonna play. Like we did back in the day, taking Nigga's fitted hats, mostly anything else was a plus at the time. Was knocking unexpected people out for years, unfortunetly, we preyed on our own kind daily. So, yes, evolution is taking place if I can change, we all can. You can do whatever you want, look at me, ex-gang banger, turned inventor and best selling author, at least, those are my goals.

CHAPTER 10

Be leaders, not followers people, please. No disrespect to anyone who don't agree, but view these photos and get a look at the real enemies I believe they aren't human. If you have seen the movie predator, they are similar in existence y'all, hard to see, but these on camera, not to often seen, with the human eyes. So yes, cell phone camera's broke this case wide open. Gave me the real evidence, not image effects, tricks of light, not imagination, as the guy from the pentagon has said, to keep the truth from the people. He also said they were interesting. The story and photos as he said back last year, when I sent them 30 of my best photo's, after Mufon contacted them and they said, when I ask for them to send my photo's back, I could not have them back, said it all for me. Plus when I met with the head of Mufon in Md., he saw just one of my best photo's I took on May 19, 2012. He said he would send one of his best guys out to see me, to check the photo's and my story. Also, told me if rea they may be worth lots of money. They are real and I've received nothing from them yet, but Mufon, was a great organization, at one time. Them, and government working together kinda knocks their credibility. I mean government says they don,t investigate UFO,s and sightings, right, but they work with Mufon on real cases, like mine. My life so far has been extra ordinary. Like one time back in 1995, the hood was in a bad way. All of what I felt, I helped put together in building clout and respect was going to be for nothing. I was facing mad jail time for serving an undercover police on Saverly and Bloomingdale, and my women was acting brand new when we had been together since 87, off and on, my rent, BGE, and lawyer fees were due, and cash was slowing down. I had pretty much pawned a lot of my gold and diamond rings because I had like a dumb in our mob inner circle crew IOPP move in with me. My mistake. Everything went to shit. This nigger, did not want to pay his way on bills, my cats ripped up his living room set, so he got pissed off at me. Yeah, I had like two fourteen carat gold rings for every finger, twice on both hands, a Gucci link chain, a herring bone necklace thick as Tammy, my home girl a... A Gucci fourteen carat link watch, that I had found, which was real and appraised for around $1000.00 at the time. I had a 14 carat gold Fugero chain and bracelet as well. I moved into my first two bedroom apartment in 91. Seton Park. After staying with my brother Rahoo's family, who lived on our block, our block. I attended to grandma Bailey those couple of years. She had cancer and went to dialysis a couple of times a week.I slept on the couch downstairs and she was in the dining room. I miss her. She gave my Aunt Yvonne the yes on letting me stay there. My step dad and I, Mr. Charles A.

Taylor Sr., was at each others throat cause he knew I wasn't doing right. Both of my parents worked and did not want me going astray. Did what they could for me and my brother, Charles Jr. We both wanted more from the streets of West Eglestone. We had lots of women too. I've never really discussed UFO's with him either. Oh yeah, in 95, I went through some real life threatening warning. Now the whole time I hustled, I made sure I bought at least three bags of weed a day, to supply my habit. My girl, smoked also. So I was going through something in the summer of 95. The mistakes I almost made then, I hope still don't cost me. I stayed up for seven straight days. I got about 7 hours of sleep after. I dreamt and watched my own murder taken place at 3603 Bowers Ave. Apt D. My girl and I was beefing because she found out that I might be the dad of my homey Rahoo's peoples kid. I think September 13, 1993 was the year she was born and her name is Paris. I was facing 25 years in jail at that time, locked up on three felony charges. No bail. Plus I had felonies I was waiting to go to court for. I had possession with intent to distribute, manufacture, conspiracy. I had another possession with intent, assault with intent to murder and some more s..., all at the same time.

Now these were my first major drug charges after making over $150,000.00 on the streets, maybe more, no weight sales, just hand to hand over a period of time. You got to remember economy was booming in the mid and late 80's and 90's during my run. Now it all started in or around 83-99. If I was home from jail, I was on the block, I mean, we were in 86 Benze's in 86. My man had his own Maxima in 87 off of street money. We all got it in. All three facites of our MOB you can't catch us all up together at the same time because we did not bring all our cash back to the same table, unless it was time. Then we made adjustments to make sure we won. Now in 95, I _____ went down L. and Payson, grabbed three pink bags from a little nigga I went tto High School with, from E.A.Edmondson Ave. He and like seven other people came up to me and was like du-du. I got the bomb G. I waived the others off and grabbed three of them. Now I don't know if it was laced, super strong, or if he purposely gave me some bull s###, because when we both went to South Brook in the 90's, we would chase him and hold him on the third floor balcony by his legs, upside down, until he cried sometimes. So maybe he thought he would get me back, but anyway, I took them, went home and smoked one blunt out of it. Not even the whole thing. Instantly, my throat stated hurting and I felt crazier, cause I was already on edge. I started feeling anxious, paranoid, ready, and violatal all at the same time. I went off on everyone around me. I u•usted no one at the time and here's why. I had a dream, when I was outside of my body, watching someone with a key to my crib, come in, come up the stairs, go to my extra room, and normally I would be in the main bedroom with my girl, but as I said, we were beefing, so whoever it was, I trusted with my life fully. I was sleep in the dream as I watched them open my door, where I slept on the bed, they pulled out a silver gun, shot me in the head first then they hit me about 7 times to my body. They wanted me dead for some reason, who knows why?but I felt every shot that hit me. I did not stand a chance and I stayed ready for war. After the last shot, I woke slam up, instantly, and I have not slept right since and don't mind sleeping on a couch today. This, I believe, was going to happen at that time, cause I knew too much about our hood to be off in anyway and niggas feel uncomfortable with it. I mean possible serial killers come in every hood, people. They are raised by a different set of rules. It's passed down from generation to generation. That's why most women go hard, look for a go harder man to create a kid killer to help protect the parents.Through kids, are by design, not coincidence, when both father and son end up in jail or dead. Big trouble, or even if they both do the right thing, it's genetics. It's with a lot of cases. I know of with my research. Yeah, that was the weirdest thing I've ever gone through because at that time, I needed a break from everybody, not a breakdown, just a break. It made me even stronger in life by going through it. The kid that sold me the herb that day, got robbed and killed later. He got shot in the head. Lil Juan, R.I.P. Now I've been present at the time of deaths to many people I've known and loved as well. I've gone to see homies from the past one day to find out the next day they were killed. Several times this has happened and I can't explain it or go into details. I was retired since 99, so I haven't really been out there lately. Not since the knockers told me to move shop. That's undercover police team. After they arrested me several times in 99, 7 times over less than a year. One of them said that our block belonged to them niggas telling. I got 4 years out of those charges, won't ever go back to making money there again, I don't think. Went in 99, came home in 02.

While in jail, I spent my time mostly out W.C.I., with real gangsters, o.k? Gangs like 276, which I felt I was a part of anyway, because I knew the Luitenant Commander since 1980. Adolf, Mr. Hester, or Lil Edgar, my mentor, my friend, my blood, was killed in 2011 or 2012. Shot to death on Pressman playground. R.I.P. I love you my OOG. He put mad work in a real legend from block. Helped keep me protected for years, but almost allowed me to be killed in 1981 by another go hard nigga at the time, Lil Mike Mike from North & Middletown. I wasn't on their level yet, but bAðh?ÍY) I was well on my way after not backing down from their aggression toward me and my younger mob at the time. There were about five of us walking up Normount and they kept walking, but something told me to <u>listen</u> Adolf called us coons, I heard him, but didn't really know him like that to allow that. So I said, "who y'all niggas talking to." I told them, " neither one of y'all talking to me." Lil Mike started pulling out the 25 auto he had in his pocket and shoot me in the head. He told Adolf and Adolf said no, I like shorty, talking about me and my heart. I've been in ever since, ok. Real business. Regardless of whatever, we were real gangland, way beffore they started making money off the glamour of gangbanging. They were banging in my hood since the 50's. We had people in place for whatever, all and any situation came up. Now this book and knowledge of understanding the hood life should enlighten y'all. Evolution is the goal. These beings are not waiting for Barack O' Bama, yeah,At time's I wrote it, to expose the truth to the people. The beings are making direct contact with those who are ready to receive the message. What I have is similar to what we do to our own kids. We give them coloring books and crossword puzzles to increase their brain power young. These aliens are doing the same thing with us. Crop circles, photos, and videos of them.

S..., sometimes, I think they use even the skies as a cosmic blackboard as a way to send the info in the clouds. So my photos of these amazing looking creatures, are showing them for one observing me, and doing things like inculating in the bushes. Releasing some type of alien radiation on film. I've seen Bigfoot, blue in color, looking at some type of Dogman being, fighting another creature. I've seen human looking beings in space suits with claws in my woods. I've seen colored copies of my dog standing against a tree watching me take the photo. Now I got all this film, while not seeing them with the naked eye all the time.

A lot of the UFO's, I have seen with the naked eye. I'm calling this the greatest UFO sighting ever captured on film. I've had hundreds of UFO sightings in my life, but on May 19, 2012, something happened that allowed me to put some of it together. I was already in a UFO ready made day, perfect day for a sighting. As I went outside, to post up, hoping I would see something, I noticed my cell phone was very low on energy. I put it on the charger and went outside around 5:15pm. As I got to the green box, about 30 feet from my house, I had a real bad feeling, cause my daughter Alicia, had to move back in with me and my wife. But I had a good feeling about seeing a UFO. I was out there for about 20 min. just looking in the sky and my woods. Now people this really happened. I have the photos to prove this. Suddenly, I saw a blue light coming up from the back part of the parking lot from the woods. It's brood daylight and I can clearly see the blue and white lights coming towards my direction. Soon as I realized what it was, I ran into my house like I was running from the police. I don't even remember putting a key in the door to get in, but I do remember my wife and daughter saying, "what the f... is wrong with you," as I burst through the door, screaming, "there is a real UFO outside." Grabbing my cell phone from my charger, ripping it almost out of the wall, I ran out the door hoping to my motherf...... surprise, it was still there 10 ft away from my door. I took a couple of steps towards it, and started taking photos with my Sprint 2700 SCP cell phone camera. I took a couple of shots before it disappeared in the woods across from me landing. Now I'm going nuts. I runs over to see if anyone else saw this even though the people upstairs and downstairs were having a cook out, they saw me, but acted as if they didn't see nothin. As I looked at the white, blue and black swirling objects on film, I can clearly see it. Then like another 10 min. later, another UFO appeared flying very low, tree top level, out of the area where the first one came from. Ok, now I'm up jumping around trying to get this on film for the world to see. I did just that. I have it on film for the world to see and I did. I have it on film too, flying through the bush line, towards power lines, across the highway behind my house. But this time, someone else sees me. He comes up and asks me, "what are you doing homie." I didn't know him, but I said, "taking pictures of UFO's that are out here flying around." He said, "no way. I was wondering if they were real." Then I showed him what I was seeing and he saw it as well. His name was T.C. and his aunt lived around the front of the complex. He is still my homie to this day and will confirm what took place during the day and time. Now just 10 min. after that, a third UFO appeared as well. This UFO did something different, it lowly flew until it rose to about 5,000 ft. in the air. It rose above the cookout crowd and hovered for about 2 mins., then it disappeared and reappeared with a fusive type thing flying around it. It did that and then shot up and disappeared. Now sadly, the photos I took of the third UFO, did not turn out, but the first two, I clearly got on film. This all took place in West Eglestone County, Woodlawn, Md, May 19, 2012. Real talk.

After experiencing that, I started looking at the photos constantly and everyday. I could and discovered the answer to human kinds biggest questions which are we, or are we not alone. I have the proof on photos and videos. UFO's and I have both aliens and UFO's. I've been investigating the subject for 33 yrs. My family and I have the evidence proving this. Thousands of photo's. I will come and check y'all hoods as well for a fee. $100.00 per person, per visit, to explore them. My info is real. This book is 95% non-fiction. 5% fiction only to protect people and their families from feds. I got loved for all those mentioned in this book. So I don't forget, one to all y'all from the Du Duster. Ok. My third eye was improved on the streets of West Eglestone, Saverly & Middletown, Bloomindale, Poplar Grove, north & Middletown, The Village, The Projects, Cherry Hill, Westport, Chedrick, Park Heights, Pigtown, C.B.S., Sandtown, EA., The Junction, Baker & Rosedale, Pressman, Ellamont, Brighton, Normount Ave., North & Pulaski, The Bridge, West Mount, Murphy Homes, Lexington Terrece, and all the rest as well. Prepare people, there is a much greater force amongst us. We must prepare for them, not each other. Time for peace y'all, FOG. The first organized team for peace period. All we want is to get money and we have no enemies in the field. We want to keep it that way. We will make cash payments to MOB. And to assure peace. Some of the proceeds will go to all those whom I mention at rate I approve. This is all about the truth. How it go down in Eglestone, o.k. All gangsters everywhere, please stop the killing soldiers. We may need to fight a war for our survival as a race. No bull s... If you got beef, then fight and may the best man win and the looser, just lose, but we all stay alive, we all win, o.k. Thank you all for the support. Hopefully, I can tell more in Part II. God willing. I have enough info for three more books, three movies and some more as well.

THE END

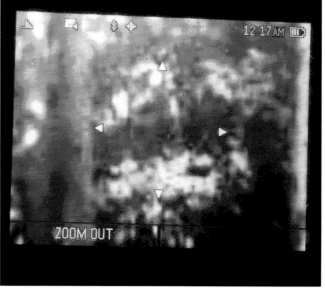

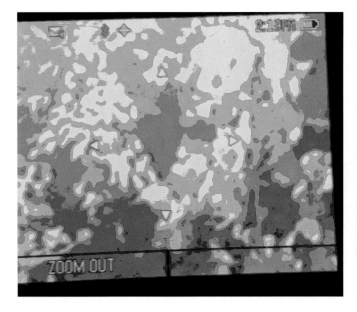

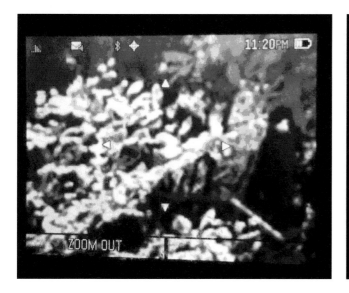
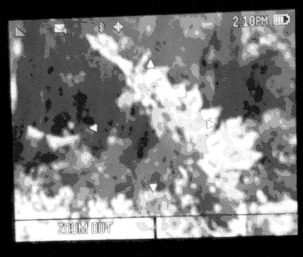

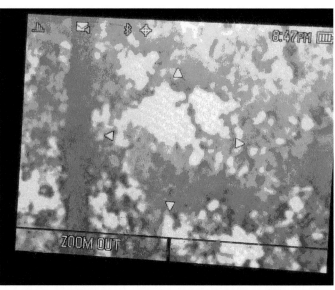

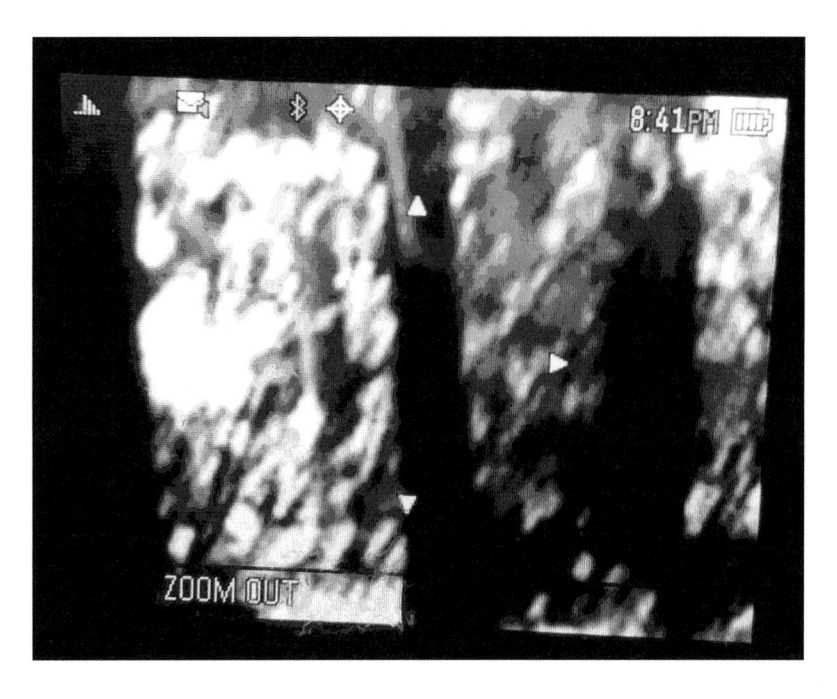

Printed in the United States
By Bookmasters